Design for Sustainable Inclusion

Joy Goodman-Deane · Hua Dong ·
Ann Heylighen · Jonathan Lazar · John Clarkson
Editors

Design for Sustainable Inclusion

CWUAAT 2023

Springer

Editors
Joy Goodman-Deane
Department of Engineering
University of Cambridge
Cambridge, UK

Hua Dong
Brunel Design School
Brunel University London
Uxbridge, UK

Ann Heylighen
Architecture
KU Leuven
Leuven, Belgium

Jonathan Lazar
Maryland Institute for Digital Accessibility
University of Maryland
College Park, MD, USA

John Clarkson
Department of Engineering
University of Cambridge
Cambridge, UK

ISBN 978-3-031-28527-1 ISBN 978-3-031-28528-8 (eBook)
https://doi.org/10.1007/978-3-031-28528-8

This Springer imprint is published by the registered company Springer Nature Switzerland AG
The registered company address is: Gewerbestrasse 11, 6330 Cham, Switzerland

Preface

The Cambridge Workshop on Universal Access and Assistive Technology (CWUAAT) series has hosted the multifaceted dialogue on design for inclusion since 2002, involving disciplines such as computer science, engineering, design, architecture, ergonomics and human factors, policy and gerontology. The workshops were held every two years at Fitzwilliam College in Cambridge University until they were disrupted by the COVID-19 pandemic in 2020. This book arises from the first CWUAAT workshop after this disruption, which was held in St Catharine's College, Cambridge, in March 2023.

The conference theme is *Design for Sustainable Inclusion*, which was inspired and informed by the United Nations Sustainable Development Goals (available at: https://sdgs.un.org/goals). These include, among others, 'good health and well-being', 'reduced inequalities' and 'sustainable cities and communities'. Addressing this challenge requires a cross-disciplinary approach, and CWUAAT 2023 invited participants from a wide variety of disciplines to discuss the implications of design for sustainable inclusion openly, critically, rigorously and imaginatively.

This book contains the papers selected for oral presentation at the conference. These were selected by peer review and carried out by an international panel of currently active researchers. The chapters within the book provide a unique insight into current national and international research in the fields of inclusive design, universal access and assistive and rehabilitative technology. The main sections of the book reflect the following key themes:

I Understanding people
Understanding the characteristics and needs of the end-users of products and services is an essential foundation for designing inclusively.

II Designing for an ageing population
The world's ageing population presents key challenges and opportunities for design. Inclusive design methods are well placed to meet these challenges and create appropriate, usable and inclusive solutions.

III Inclusive built environments
Inclusive built environments have great potential to improve independence and inclusion within communities.

IV Healthcare
Inclusive design is essential to ensure that no one is excluded from essential healthcare services.

V Assistive technology
Alongside inclusive design, there remains a need to develop solutions specifically to support individuals with disabilities.

VI Design methods
Good methods for supporting inclusive design and the design of assistive technology are essential for helping designers to put these into practice.

VII Education and training
This section examines how to educate designers in inclusive design and how to make education itself more inclusive.

We would like to thank all the authors and researchers who have contributed to CWUAAT 2023 and to the preparation of this book. We would also like to thank all the reviewers, who helped to ensure the high quality of the content. We are particularly grateful to Anthea Maybury and Mari Huhtala who played a key role in readying the book for publication.

March 2023

Joy Goodman-Deane
Hua Dong
Ann Heylighen
Jonathan Lazar
John Clarkson

Contents

List of Contributors

A. Balmaseda Dominguez Faculty of Architecture and Urban Planning, University of Stuttgart, Stadtplanung und Entwerfen (SuE), Stuttgart, Germany;
Department of Architecture, Università degli Studi Roma Tre, Rome, Italy

M. Bénéjean University Rennes 2 – PREFICS, Rennes, France

P. R. Bohn Federal University of Health Sciences of Porto Alegre (UFCSPA), Porto Alegre, Brazil

C. de Bont School of Design and Creative Arts, Loughborough University, Loughborough, UK

M. D. Bradley Engineering Design Centre, Department of Engineering, University of Cambridge, Cambridge, UK

A. M. Brock ENAC, Université Toulouse, Toulouse, France

N. Carmona Institute for Sustainability Science and Technology, Universitat Politècnica de Catalunya, Barcelona, Spain;
Innovation Consultancy Department, Fundació Eurecat, Barcelona, Spain

P. Carnemolla Faculty of Design Architecture and Building, School of Built Environment, University of Technology Sydney, Ultimo, Australia

E. Ceretto Fans4All Association, Paris, France

J. Chakraborty Department Computer and Information Sciences, Towson University, Towson, USA

C. Y. Choi Royal College of Art, Design Products, London, UK

P. J. Clarkson Engineering Design Centre, Department of Engineering, University of Cambridge, Cambridge, UK

H. Dong College of Engineering, Design and Physical Sciences, Brunel Design School, Brunel University London, London, UK

J. J. Dudley Engineering Design Centre, Department of Engineering, University of Cambridge, Cambridge, UK

H. Corderoy Du Tiers Fans4All Association, Paris, France

J. H. Feng Department of Computer and Information Sciences, Towson University, and On Sabbatical, Trace Research and Development Center, College of Information Studies (iSchool), University of Maryland, College Park, USA

M. Forbes Department Computer and Information Sciences, Towson University, Towson, USA

List of Contributors

A. Balmaseda Dominguez Faculty of Architecture and Urban Planning, University of Stuttgart, Stadtplanung und Entwerfen (SuE), Stuttgart, Germany;
Department of Architecture, Università degli Studi Roma Tre, Rome, Italy

M. Bénéjean University Rennes 2 – PREFICS, Rennes, France

P. R. Bohn Federal University of Health Sciences of Porto Alegre (UFCSPA), Porto Alegre, Brazil

C. de Bont School of Design and Creative Arts, Loughborough University, Loughborough, UK

M. D. Bradley Engineering Design Centre, Department of Engineering, University of Cambridge, Cambridge, UK

A. M. Brock ENAC, Université Toulouse, Toulouse, France

N. Carmona Institute for Sustainability Science and Technology, Universitat Politècnica de Catalunya, Barcelona, Spain;
Innovation Consultancy Department, Fundació Eurecat, Barcelona, Spain

P. Carnemolla Faculty of Design Architecture and Building, School of Built Environment, University of Technology Sydney, Ultimo, Australia

E. Ceretto Fans4All Association, Paris, France

J. Chakraborty Department Computer and Information Sciences, Towson University, Towson, USA

C. Y. Choi Royal College of Art, Design Products, London, UK

P. J. Clarkson Engineering Design Centre, Department of Engineering, University of Cambridge, Cambridge, UK

H. Dong College of Engineering, Design and Physical Sciences, Brunel Design School, Brunel University London, London, UK

J. J. Dudley Engineering Design Centre, Department of Engineering, University of Cambridge, Cambridge, UK

H. Corderoy Du Tiers Fans4All Association, Paris, France

J. H. Feng Department of Computer and Information Sciences, Towson University, and On Sabbatical, Trace Research and Development Center, College of Information Studies (iSchool), University of Maryland, College Park, USA

M. Forbes Department Computer and Information Sciences, Towson University, Towson, USA

A. Forsythe-Korzeniewicz Self-Advocate, Baltimore, USA

V. Garaj Brunel Design School, Brunel University London, Uxbridge, UK

T. Golgolnia Chairs of Social and Health Care Buildings and Design, Faculty of Architecture, Technische Universitaet Dresden, Dresden, Germany

J. A. Goodman-Deane Engineering Design Centre, Department of Engineering, University of Cambridge, Cambridge, UK

R. Griffiths Cartwright Pickard Architects, London, UK

J. Guan School of Design and Creative Arts, Loughborough University, Loughborough, UK

E. Healy Interaction Design Centre, Department of Computer Science and Information Systems, University of Limerick, Limerick, Ireland

F. Hémont University Rennes 2 – PREFICS, Rennes, France

A. Heylighen Department of Architecture, Research[x]Design, KU Leuven, Leuven, Belgium

I. M. Hosking Engineering Design Centre, Department of Engineering, University of Cambridge, Cambridge, UK

A. Isha Department of Architecture and Planning, Indian Institute of Technology-Roorkee, Roorkee, India

A. Kaniadakis College of Engineering, Design and Physical Sciences, Computer Sciences, Brunel University London, Uxbridge, UK

M. Kevdzija Institute of Architecture and Design, Faculty of Architecture and Planning, TU Wien, Vienna, Austria

E. Konig Federal University of Health Sciences of Porto Alegre (UFCSPA), Porto Alegre, Brazil

P. O. Kristensson Engineering Design Centre, Department of Engineering, University of Cambridge, Cambridge, UK

J. Lazar Maryland Institute for Digital Accessibility, University of Maryland, College Park, USA

B. Lazzarini Institute for Sustainability Science and Technology, Universitat Politècnica de Catalunya, Barcelona, Spain

C. S. Libânio Federal University of Health Sciences of Porto Alegre (UFCSPA), Porto Alegre, Brazil

C. Lounis SII Research & ISAE-SUPAERO, Université de Toulouse, Toulouse, France

J. L. Madras Fans4All Association, Paris, France

A. Manohar College of Engineering, Design and Physical Sciences, Brunel Design School, Brunel University London, London, UK

G. Marquardt Chairs of Social and Health Care Buildings and Design, Faculty of Architecture, Technische Universitaet Dresden, Dresden, Germany

P. Marrast CERTOP, UMR 5044, Université de Toulouse, Toulouse, France

C. McGinley The Helen Hamlyn Centre for Design, Royal College of Art, London, UK

Q. Ni College of Engineering, Design and Physical Sciences, Brunel Design School, Brunel University London, London, UK

H. K. Oswal Science and Math Institute, Tacoma, WA, USA

S. K. Oswal Center for Research and Education on Accessible Technology and Experiences, School of Interdisciplinary Arts and Sciences, University of Washington, Seattle/Tacoma, WA, USA

E. Peyrard i3-CRG, UMR 9217 CNRS, École Polytechnique, Palaiseau, France

F. Pour-Rahimian Centre of Construction Innovation and Research, School of Computing, Engineering and Digital Technologies, Teesside University, Middlesbrough, UK

G. Raheja Department of Architecture and Planning, Indian Institute of Technology-Roorkee, Roorkee, India

C. Resendiz-Villasenor Centre of Construction Innovation and Research, School of Computing, Engineering and Digital Technologies, Teesside University, Middlesbrough, UK

E. Roca Institute for Sustainability Science and Technology, Universitat Politècnica de Catalunya, Barcelona, Spain

D. Rosen Institute of High Performance Computing, A*STAR, Singapore, Singapore

J. Sturge Department of Design, Production and Management, Faculty of Engineering Technology, University of Twente, Enschede, The Netherlands

L. M. Tonetto National Council for Scientific and Technological Development, Brasília, Brazil

P. Tutenel Department of Architecture, Research[x]Design, KU Leuven, Leuven, Belgium

S. D. Waller Engineering Design Centre, Department of Engineering, University of Cambridge, Cambridge, UK

R. E. Wood Maryland Institute for Digital Accessibility, University of Maryland, College Park, USA

B. Yang Engineering Design Centre, Department of Engineering, University of Cambridge, Cambridge, UK

L. Yin Brunel Design School, Brunel University London, Uxbridge, UK

C. Zecca The Helen Hamlyn Centre for Design, Royal College of Art, London, UK

E. Zitkus School of Design and Creative Arts, Loughborough University, Loughborough, UK

Understanding People

Understanding Real People: Going Beyond Single Severe Capability Loss

S. D. Waller(✉), J. A. Goodman-Deane, M. D. Bradley, I. M. Hosking, and P. J. Clarkson

Engineering Design Centre, Department of Engineering, University of Cambridge, Cambridge, UK

{sdw32,jag76,mdb54,imh29,pjc10}@cam.ac.uk

Abstract. Some approaches to representing the diverse range of capabilities within the population focus on representing those with severe losses in a single capability, such as vision or mobility. This approach follows the simplistic view that people are either able-bodied, or have a single severe capability limitation. This paper challenges this view using data from a population representative survey of the 2019 adult population in Germany. This survey asked participants whether they had any limitations in daily activities because of their vision, hearing, memory, concentration, hands, reach or mobility. Out of 1002 valid responses, 158 participants reported that something was 'very limited'. However, 145 of these participants (92%) had a limitation in more than one capability category. Therefore, representing the diverse range of capabilities within the population necessarily requires describing people with co-occurring limitations. The full set of capability limitations for an individual is defined herein as their 'capability fingerprint'. Within the German survey, the top 16 capability fingerprints represented 80% of the survey participants. Presenting the most prevalent capability fingerprints is advocated as a better approach for understanding the diversity of the population. These capability fingerprints could be used to develop population-representative personas, or to better understand the sample that took part in small-scale user research.

1 Introduction

Inclusive design is 'the design of mainstream products and/or services that are accessible to, and usable by, as many people as reasonably possible' (BSI 2005). Inclusive design is about designing for the 'whole population' and is therefore necessarily underpinned by approaches for understanding and representing the diverse range of capabilities within the population (Clarkson et al. 2007).

The Microsoft Inclusive Design Toolkit (Shum et al. 2016) advocates 'The Persona Spectrum'. This includes someone with one arm, someone who is blind, someone who is deaf, and people with several other temporary and situational impairments.

Schulz and Fuglerud (2012) claim that a good start for creating personas with disabilities is to create one persona for each of the 4 main groups of disabilities (vision, hearing, movement, cognitive), and then complement this with one or two elderly personas with

J. Goodman-Deane et al. (Eds.): CWUAAT 2023, *Design for Sustainable Inclusion*, pp. 3–12, 2023.
https://doi.org/10.1007/978-3-031-28528-8_1

combinations of milder versions of these impairments, although the paper does not provide any prevalence-based rationale by which combinations should be chosen.

Both of these approaches to representing the diversity of capabilities in the population follow the simplistic view that most people are either able-bodied or have a single severe capability limitation.

This paper examines the extent to which approaches like these are corroborated by survey data, by examining the prevalence of singular and co-occurring limitations. It presents data from the DIGNITY digital exclusion survey, which was conducted in Germany in 2019 (Goodman-Deane and Waller 2022). Some of the questions within this survey were about capability loss, and these responses were analysed to consider the prevalence of those with just one kind of capability limitation, as compared to those who have multiple co-occurring limitations.

Subsequently, the results are presented for each capability category in turn, to show the relative prevalence of the different kinds of limitations. The results are briefly compared to the prevalence reported from some other surveys. Finally, a new approach is offered for presenting the prevalence of co-occurring capability limitations, which is named here as 'frequency counting the capability fingerprints'. This paper briefly describes this approach, then presents the most prevalent capability fingerprints, and then discusses using these fingerprints to understand the kinds of people that didn't take part in opportunistically sampled small-scale user research. Finally, the paper presents the limitations of this analysis, followed by the conclusions and further work.

2 Survey Method and Analysis

The DIGNITY surveys (Goodman-Deane and Waller 2022) were conducted in five European countries and regions, to understand the factors that might exclude people from public transport services, with a specific focus on digital exclusion. The results presented in this paper are based on the DIGNITY survey that was conducted in Germany (hereafter referred to as 'the survey'), because it had the largest sample and the most reliable sampling method, which was based on adults aged 16+ living in private households. The survey dataset is available open access (Goodman-Deane et al. 2022b). More information on the survey, its weighting, sample and results can be found in Goodman-Deane et al. (2022a).

This paper focuses on the survey's self-report questions about the impact of capability loss on everyday life, which were presented to the participants as follows (translated into German): 'Now there are a few questions about how much your capabilities affect your daily life. In these questions, we're interested in your real-life experience e.g. while using any glasses or other aids you normally use. To what extent, if at all, are you limited in your daily activities because of **your eyesight**'. The possible answers were 'not at all limited', 'somewhat limited', 'very limited' and 'prefer not to answer' (translated into German).

After this question, the survey asked the same question about: '... **your hearing** ... any problems with **your hands** ... any problems with **your mobility**... any problems with **reaching your arms** above your head or out to the sides... any difficulties with your **memory or concentration**' (translated into German). In order to simplify the results, the responses for 'hands' and 'reaching your arms' were combined, so the maximum reported limitation in these two source questions was used to give an overall limitation for hands/reach.

The dataset was weighted for age, gender and region. All the results presented in this paper used the weighted dataset, and all values of N are rounded to the nearest whole number. 'Prefer not to answer' was treated as missing data in the analysis. Of the 1010 participants in the survey, 8 had missing data for one or more of the capability questions, giving N = 1002 valid responses.

3 Visualising the Prevalence of Co-occurring Capability Losses

Benktzon (1993) introduced a segmented pyramid as a simple model for describing the variation of capability within the whole population, with the bottom segment representing those who are fully able, a middle segment representing those with minor capability loss, and the top segment representing those with severe capability loss.

Waller et al. (2010) refined this model to show additionally the prevalence of singular vs co-occurring losses. Figure 1 applies the same approach to the DIGNITY German survey. The segments of this pyramid correspond to the survey responses of 'not at all limited', 'somewhat limited' and 'very limited'.

Figure 1 shows that less than half of the survey participants (40.1%) reported 'not at all limited' in all 5 categories, and more than half (59.9%) had some kind of limitation. The survey intentionally asked participants whether they had any kind of limitation in their daily activities, whereas disability-based surveys (e.g. Grundy et al. 1999) only consider 'long-term' impairments that are severe enough to 'hinder full and effective participation in society', and therefore be considered as a disability (United Nations 2006). Broadening the scope of the limitations that are examined within the survey has considerably decreased the proportion of people who are 'fully able', down from 81.5% (Grundy et al. 1999; Waller et al. 2010) to 40.1% in the DIGNITY German survey.

The co-occurrence results will now be considered in more detail. Out of 1002 valid respondents, 158 participants reported that something was 'very limited', yet only 13 out of these 158 participants (8%) had a single capability that was very limited, with no other kinds of limitations. Comparatively, 145 of these participants (92%) had multiple co-occurring limitations (of which at least one was very limited).

So, as a first order approximation, according to the survey, if one wanted to create population-representative personas with severe kinds of capability loss, about 90% of them should have co-occurring losses, and about 10% of them should have a single severe loss. This radically contrasts with the disability-based persona sets introduced in the introduction (Schulz and Fuglerud 2012; Shum et al. 2016), where between 0 and about 30% of the personas had co-occurring losses.

Figure 2 shows an alternative way of presenting the prevalence of co-occurring limitations. This stacked bar chart shows the percentage of the sample who had exactly 0, 1, 2, 3, 4 or 5 limitations (counting somewhat limited or worse). The prevalence is shown separately for age 16–64 and age 65+. It is well-known that the prevalence of limitations increases with age (e.g. Clarkson et al. 2007), so it's no surprise that the vast majority of people with 4 or 5 limitations are aged over 65.

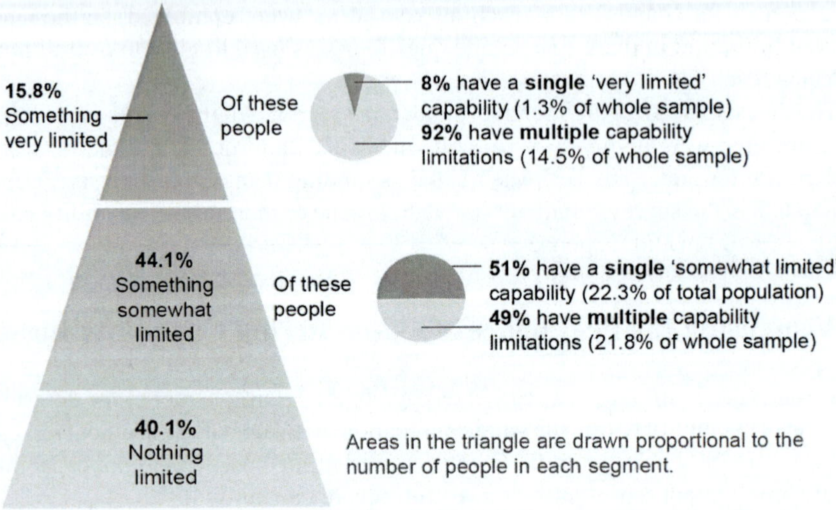

Fig. 1. This segmented pyramid provides a simple model for understanding the prevalence of capability limitations in the population, and the prevalence of co-occurring limitations. Areas in the triangle are proportional to the number of people in each segment.

Furthermore, it is no surprise that the vast majority of people with 0 or 1 limitation are aged 16–64. However, it's interesting to see that the people with 2 or 3 limitations are close to being evenly split between those aged 16–64 and aged 65+.

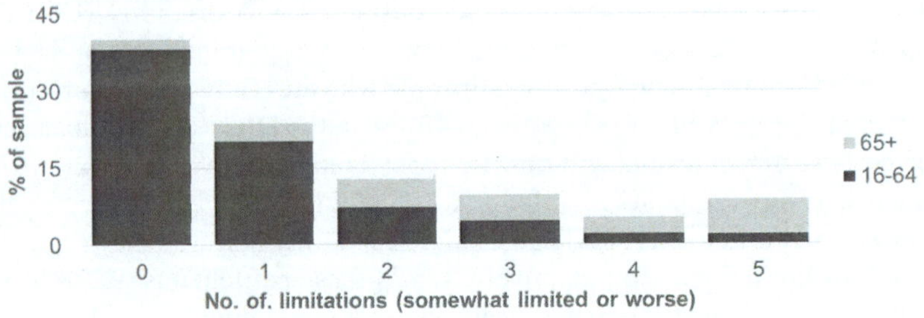

Fig. 2. The prevalence of different numbers of limitations, shown separately for those aged 16–64, and those aged 65+

However, knowing that co-occurring capability loss exists is not enough to fulfil the aim of inclusive design to 'design for the whole population'. To achieve this, it is necessary to know exactly which kinds of capability losses are more prevalent than others, and exactly which kinds of co-occurrence are more common than others.

Figure 3 examines each capability type in turn, and shows the prevalence of limitations in that capability, broken down according to 'somewhat limited' or 'very limited'. Figure 3 shows that vision was the most prevalent cause of limitations. Just over half of the sample (51.3%) reported some kind of limitation in this category, approximately twice as much as any of the other categories. Comparatively, the same set of activity limitations questions were asked in the UK 2010 Towards Better Design survey, with 362 participants (Clarkson et al. 2012). In this case 10.5% of the sample indicated some kind of limitations due to vision, 5.6% due to hearing, 15.2% due to memory/concentration, 14.3% due to hands/arms, and 15.5% due to mobility.

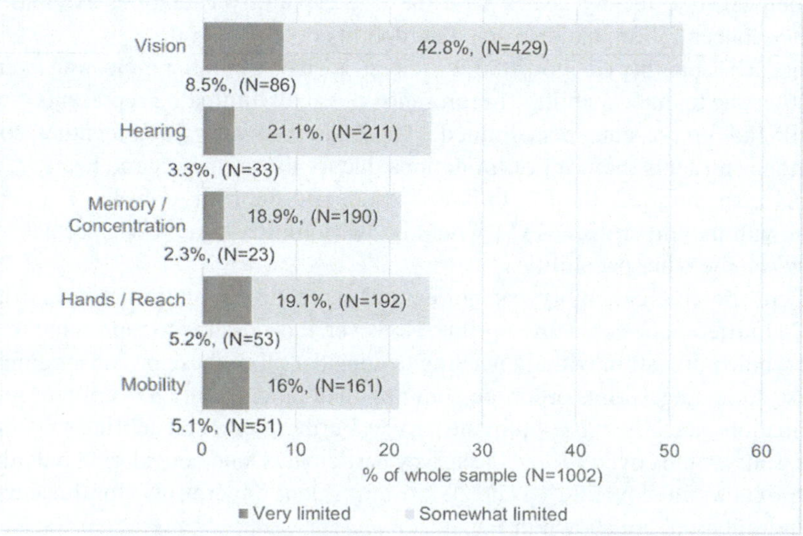

Fig. 3. Prevalence of limitations, considering each capability separately

The same questions were also asked in the Italian version of the DIGNITY survey (Goodman Deane et al. 2022.c), with 1002 participants, and in this case 29% reported some kind of limitation due to vision, 15.7% due to hearing, 14.2% due to memory/concentration, 18.6% due to hands/arms, and 15.1% due to mobility.

This preliminary investigation reveals that, across all the categories, the German survey participants reported more limitations than the Italian and UK participants, and especially so for vision. The authors can only speculate some possible reasons for this:

- the meaning of ‘somewhat limited’ and ‘very limited’ are subtly different in the different languages;
- the actual prevalence of impairments in Germany is higher, or the effectiveness of assistive devices in mitigating these is lower;
- for an equivalent level of functional capacity, the German respondents were more likely to self-report activity limitations. This could perhaps be because the things that the Germans typically interact with require comparatively higher levels of capability.

The authors acknowledge that the higher prevalence of limitations reported in the German survey will inevitably lead to higher levels of co-occurrence being reported. However, it is outside of the scope of this paper to perform a detailed comparison of the prevalence of co-occurring limitations that were reported in different countries at different times.

4 Capability Fingerprints

The most prevalent combinations of co-occurrence that exist within the DIGNITY German survey will now be presented using the concept of 'capability fingerprints'. A capability fingerprint uniquely identifies the full set of limitations experienced by a participant, due to any issues with the five capability categories (vision, hearing, memory/concentration, hands/reach and mobility).

For each capability category, the participant is given a number indicating their level of limitation due to that capability: 1 represents 'not at all limited', 2 represents 'somewhat limited' and 3 represents 'very limited'. These numbers are joined together, so that the first digit represents the level of limitations due to vision, the second hearing, the third memory/concentration, the fourth hands/reach, and the fifth mobility. For example, a person with the fingerprint 31111 would be 'very limited' due to vision, but 'not at all limited' on the other capabilities.

There are five capability categories and three levels of limitation, so there are $3^5 = 243$ different possible fingerprints. However, it is possible to represent the majority of the whole population without needing to consider all of these possible combinations, because some fingerprints are more common than others. Only 97 different fingerprint combinations actually exist within the survey. Furthermore, considering only the top 36 fingerprints would cover 90% of the survey participants, and considering only the top 16 fingerprints would cover 80%. The 20 most prevalent fingerprints (for those with some kind of limitation) are shown in Fig. 4.

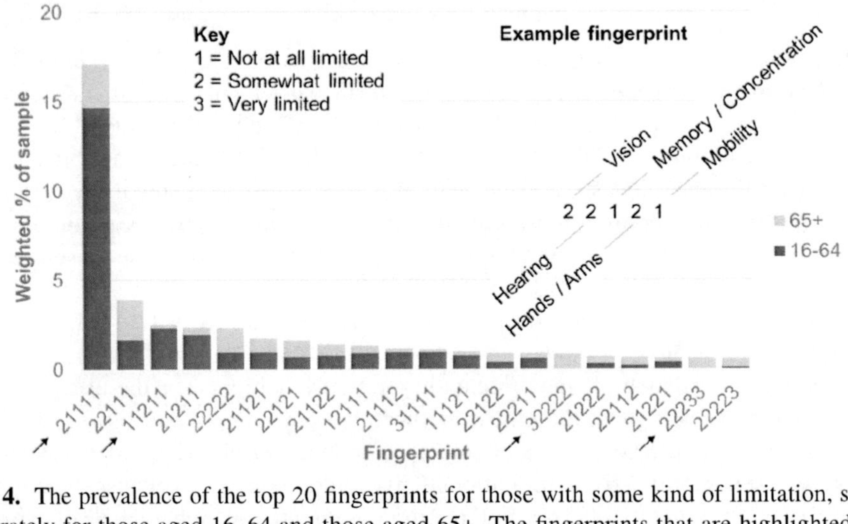

Fig. 4. The prevalence of the top 20 fingerprints for those with some kind of limitation, shown separately for those aged 16–64 and those aged 65+. The fingerprints that are highlighted with arrows are discussed further in the body of this paper.

These fingerprints are relatively straightforward to derive and frequency count, either directly within IBM's Statistical Package for Social Sciences (SPSS), or using pivot tables within Microsoft Excel. The process of frequency counting fingerprints shares many similarities with performing a cluster analysis on the dataset, and a detailed comparison of these approaches is recommended for further work.

Figure 4 shows that for some fingerprints (e.g. 21111 & 11211), the majority of people with this fingerprint are aged 16–64. For other fingerprints (e.g. 32222 & 22233), the majority of people with this fingerprint are aged 65+.

The fingerprints shown in Fig. 4 provide a rational starting point for representing the diverse range of capabilities and co-occurrences within the whole population, which goes beyond stereotypical assumptions of young people who have no limitations and older people who have some limitations. The number of fingerprints that should be presented will depend on the purpose that they will be used for, and the fingerprints will be most useful if they are derived for a specific purpose. The number of unique fingerprints that will be produced depends on:

- the survey sample that is used (e.g. Germany);
- the degree of filtering that is applied to the survey sample (e.g. age/gender, region);
- the number of capability categories that are used (e.g. mobility and hearing might not be relevant for some applications);
- the number of levels used to describe each capability category (e.g. this paper used 3 levels, but this could be simplified down to 2 levels).

The decisions made in producing the fingerprints can be iterated until the number of fingerprints is appropriate for the specific purpose that they are created for. This process could also incorporate other user attributes, like digital competencies and attitudes to technologies (Goodman Deane et al. 2021).

Furthermore, the fingerprints become extremely useful if the same questions are administered in both a nationally representative survey, and also to the people who take part in small-scale user research, which is typically opportunistically sampled. Once the fingerprints of the users that took part in the research have been derived, these can be compared to the prevalence of fingerprints from the national survey (e.g. Fig. 4), which helps to answer the following questions:

- how many people in the population have similar characteristics to each of the user-research participants?
- what percentage of the whole population is represented by the sample of users that have been recruited so far?
- what kinds of people haven't taken part in the opportunistically sampled user research?
- is the sample skewed towards those who have capability limitations, or those that have no limitations?

Petyaeva et al. (2022) used a similar approach to compare the opportunistically-sampled users that took part in user trials with a set of digital personas that were derived from cluster analysis of survey data, and further work in this area is recommended.

5 Limitations

This paper considers an approach to representing the diverse range of capabilities within the whole adult population (16+). Therefore, the results can only be derived from surveys that intend to sample the whole adult population. The Dignity German survey is one such survey with this remit, which had 1010 respondents. However, even with this sample size, only 13 respondents reported a single capability being 'very limited', with no other limitations.

Although a detailed consideration of confidence intervals is outside of the scope of this paper, the authors acknowledge that many of the frequency counts presented in this paper are too small to achieve statistical significance. Furthermore, these results are based on self-report questions regarding limitations in daily activities, and different results might be expected if performance measures were used, and/or if the questions instead asked about functional capacities (Johnson et al. 2010).

Achieving a manageable number of fingerprints necessarily involves using a small number of levels to define the level of limitation within each capability category. Therefore, each individual level covers people who might be quite different in terms of the severity of their limitation. For example, the people who are 'very limited' due to vision may include those who are partially sighted, and also those who are blind, yet the most appropriate solutions for these different levels of limitation might be very different.

This paper assumes that the intention is to understand and represent the limitations that exist within the whole population, prioritised according to the prevalence of these limitations. However, other design scenarios may have different intentions, like designing for particular subgroups of people. In this case, it might be more appropriate to generate personas directly from known users within this subgroup, using qualitative rather than quantitative methods.

Nevertheless, despite these limitations, the approach presented herein could help to give an indication of how many people in the population are 'similar to' particular known users or personas, and this information ought to be of interest. Further work is recommended to explore the use of population-based personas and known-users to bridge quantitative and qualitative methods of representing users.

6 Conclusions

This paper has examined the prevalence of capability limitations from a survey that aimed to be representative of the entire 16+ population in Germany. This survey asked people whether they had any limitations in their daily activities due to their capabilities, regardless of whether these limitations were severe enough to merit being counted as a disability.

The results from this particular survey indicate that over half of the sample self-reported some kind of limitation due to their capabilities, with vision being the most commonly reported issue. Out of 1002 valid responses, 158 participants reported that something was 'very limited', yet only 13 out of these participants (8%) reported a limitation in just one capability category, with no other kinds of limitation. This result contradicts existing approaches for representing diversity in the population, which under-emphasise the significance of co-occurring capability losses, and over-emphasise single severe kinds of capability limitations.

Capability fingerprints have been introduced as a better way of understanding the diverse range of co-occurring capability limitations. In this particular survey, 16 different capability finger prints were sufficient to represent 80% of the survey participants. Further work is recommended to apply this approach more widely.

The kinds of people that have not taken part in opportunistically sampled user research can be considered by comparing the fingerprints of those that took part in the user research with the fingerprints from a nationally representative survey, and further work in this area is recommended.

Acknowledgements. This research was done as part of the DIGNITY project which received funding from the European Union's Horizon 2020 research and innovation programme under grant agreement number 875542. The author has applied a Creative Commons Attribution (CC-BY) licence to any Author Accepted Manuscript version arising from this submission.

References

Benktzon, M.: Designing for our future selves: the Swedish experience. Appl. Ergon. **24**(1), 19–27 (1993)

BSI: BS 7000, Design management systems – Part 6: Managing inclusive design – Guide. The British Standards Institution, London, UK (2005)

Clarkson, P.J., Coleman, R., Hosking, I., Waller, S.: Inclusive design toolkit. EDC, Cambridge, UK (2007). www.inclusivedesigntoolkit.com. Accessed Sept 2022

Clarkson, P.J., Huppert, F.A., Tenneti, R., Waller, S., Goodman-Deane, J., et al.: Towards better design, 2010. [data collection]. UK Data Service, SN: 6997 (2012). https://doi.org/10.5255/UKDA-SN-6997-1

Goodman-Deane, J., Bradley, M., Waller, S., Clarkson, P.J.: Developing personas to help designers to understand digital exclusion. Proc. Des. Soc. **1**, 1203–1212 (2021). https://doi.org/10.1017/pds.2021.120

Goodman-Deane, J., Kluge, J., Roca Bosch, E., Nesterova, N., Bradley, M., et al.: Towards inclusive digital mobility services: a population perspective. Interact. Comput. **33**(4), 426–441 (2022a). https://doi.org/10.1093/iwc/iwac014

Goodman-Deane, J., Kluge, J., Waller, S., Roca Bosch, E.: DIGNITY digital exclusion dataset – Germany (2022b). https://doi.org/10.5821/data-2117-367811-1

Goodman-Deane, J., Waller, S.: D1.2 Benchmark of factors affecting use of digital products and services across Europe. DIGNITY project deliverable (2022). www.dignity-project.eu/wp-content/uploads/2022/07/Dignity-D1.2-submitted-v02.pdf. Accessed Sept 2022

Goodman-Deane, J., Waller, S., Roca Bosch, E.: DIGNITY digital exclusion dataset – Italy (2022c). https://doi.org/10.5821/data-2117-370588-1

Grundy, E., Ahlburg, D., Ali, M., Breeze, E., Sloggett, A.: Research report 94: Disability in Great Britain. Corporate Document Services, London, UK (1999)

Johnson, D., Clarkson, P.J., Huppert, F.: Capability measurement for inclusive design. J. Eng. Des. **21**(2–3), 275–288 (2010). https://doi.org/10.1080/09544820903303464

Petyaeva, A., Goodman-Deane, J., Bradley, M., Waller, S., Clarkson, J.: Improving our understanding of user trial samples using survey data. In: Lockton, D., et al. (eds.) DRS 2022: Bilbao (2022). https://doi.org/10.21606/drs.2022.676

Schulz, T., Skeide Fuglerud, K.: Creating personas with disabilities. In: Miesenberger, K., Karshmer, A., Penaz, P., Zagler, W. (eds.) ICCHP 2012. LNCS, vol. 7383, pp. 145–152. Springer, Heidelberg (2012). https://doi.org/10.1007/978-3-642-31534-3_22
Shum et al.: Microsoft inclusive design toolkit (2016). www.microsoft.com/design/inclusive. Accessed Sept 2022
United Nations: UN Convention on the Rights of Persons with Disabilities (2006). www.un.org/disabilities/documents/convention/convention_accessible_pdf.pdf. Accessed Sept 2022
Waller, S., Langdon, P., Clarkson, P.J.: Designing a more inclusive world. J. Integr. Care **18**(4), 19–25 (2010). https://doi.org/10.5042/jic.2010.0375

Inclusivity Requirements for Immersive Content Consumption in Virtual and Augmented Reality

L. Yin[1], J. J. Dudley[2], V. Garaj[1], and P. O. Kristensson[2](✉)

[1] Brunel Design School, Brunel University London, Uxbridge, UK
{lulu.yin3,vanja.garaj}@brunel.ac.uk

[2] Engineering Design Centre, Department of Engineering, University of Cambridge, Cambridge, UK
{jjd50,pok21}@cam.ac.uk

Abstract. Immersive technologies, such as virtual reality (VR) and augmented reality (AR), enable many new possibilities for audio-visual content consumption, in particular as a means of allowing for various new experiences that either mimic or surpass what can be experienced in real life. Since immersive experiences can add variety to a daily routine and thus improve quality of life, VR and AR content may be especially attractive to individuals with disabilities and older people who are often home-bound due to limited mobility. However, for VR and AR to be able to reach their full potential, it is important to ensure the design of such systems is inclusive. As an initial step in this direction, this paper provides a set of high-level user requirements from eight focus groups that involved users with a range of disabilities/capability loss types and consequent access barriers to the full enjoyment of VR and AR content and experiences. We clustered the user requirements around the participants' prioritised strategies for overcoming access barriers. These strategies are customisation, interaction, information, and adaptation for VR; and customisation, interaction, and awareness for AR. Overall, we identify several common high-level user requirements across both VR and AR, including the need to support users in fine-tuning settings and to have available a rich number of modalities to support flexible interaction.

1 Introduction

Recent hardware and software developments have resulted in an ever-expanding range of virtual reality (VR) and augmented reality (AR) headsets offering increasingly advanced capabilities to engage in immersive audio-visual content, including movies, games, and interactive remote experiences, such as virtual attendance of a guided museum tour and participation in a music concert.

Such forms of content consumption open up new possibilities for a wide range of immersive environments. However, for immersive content to be accessible to a wide range of user groups, it is necessary for system designers and developers to have an accurate understanding of users, including user groups of people with different disabilities/types of capability loss, and older people.

To address this gap in the literature, this paper reports the results of a user requirements elicitation study that consisted of eight focus groups involving users across the

J. Goodman-Deane et al. (Eds.): CWUAAT 2023, *Design for Sustainable Inclusion*, pp. 13–21, 2023.
https://doi.org/10.1007/978-3-031-28528-8_2

disability/capability loss spectrum. Our aim is to elicit high-level user requirements for inclusive content consumption that can serve as a starting point for a more refined and nuanced elicitation of requirements to cover the functions specifically related to VR and AR content consumption.

A central approach in this work is the concept of inclusive design (Clarkson et al. 2003; Keates and Clarkson 2010), which suggests we can reach much broader and more diverse user populations by specifically attempting to better understand user diversity. The consideration of inclusive content consumption in VR and AR is perhaps particularly important given the potential that such immersive experiences have to contribute to the quality of life for people with disabilities and older people (Garaj et al. 2022). As rationalised by the social model of disability (Shakespeare 2006), real-world social and physical environments can form profound constraints on disabled and older people's access to daily life by limiting their physical mobility. VR and AR environments, on the other hand, facilitate virtual mobility, and may be able to substitute for the lack of real-world access and thus improve levels of engagement with life and its quality. Recently proclaimed visions of the Metaverse (Ravenscraft 2022) and the role it may play in complementing everyday interactions in society highlight a time critical need to make immersive technologies inclusive.

So it is unsurprising that industry has formed initiatives to tackle the broader accessibility issues. Two examples of these initiatives include XR Access Initiative (2022) and XR Association (2022). Targeted academic research still remains rather limited, but initial efforts include the survey by Wong et al. (2017) of attitudes and sentiments held by users with disabilities towards AR and VR and also the work by Garaj et al. (2019) on the inclusive design of immersive reality.

2 User Requirements Study

To elicit high-level inclusivity user requirements for content consumption in VR and AR, we carried out eight focus groups (FGs), four exploring VR (FG1-4), and a further four exploring AR (FG5-8), as summarised in the table below. The focus groups were carried out remotely on Zoom and participants were asked to reflect jointly on any past experiences with VR or AR and on specific videos displayed during the session showcasing representative experiences of VR and AR content.

Participants: Focus group participants were recruited from a user panel managed by an external inclusive research and innovation consultancy. This user panel enabled efficient recruitment and facilitated stratified sampling of a wide range of capability loss types. Where possible, participants were assigned to focus groups based on their dominant capability loss. This assignment was imperfect due to scheduling constraints, and so there was some mixing of capability loss types within groups to ensure a reasonable number of participants per group. The assignment of participants to focus groups is summarised in Table 1. Some participants participated in both the VR and AR focus groups. Otherwise, all focus groups on a particular technology were unique.

Method: The focus group structure had four parts. Part 1 served as a warm-up and probed the group's general reflections on their overall prior experiences with VR or AR.

Table 1. Focus group participants summary

Technology	Group	Disability/Capability loss type							Total	Female:Male
		Sight	Hearing	Touch	Speech	Cognition	Mobility	Multiple		
VR	FG1	3	1	–	–	1	1	–	**6**	4:2
	FG2	–	–	–	–	1	3	1	**5**	5:0
	FG3	–	–	–	–	4	1	–	**5**	2:3
	FG4	1	1	–	–	–	1	1	**4**	2:2
AR	FG5	3	–	–	1	–	4	–	**8**	6:2
	FG6	–	3	–	–	2	–	–	**5**	2:3
	FG7	–	–	1	–	1	4	–	**6**	4:2
	FG8	–	–	–	–	1	5	–	**6**	5:1
Overall Total		**7**	**5**	**1**	**1**	**10**	**20**	**2**	**45**	30:15

Part 2 involved a more detailed exploration of the group's experiences in VR or AR through review of the user experience (UX) journey. The group reflected on past experiences and specific videos showcasing scenarios of (1) putting on and using relevant hardware (i.e. headsets and hand controllers for VR and tablets and smartphones for AR); (2) interacting with menus and other user interface elements; (3) the virtual content, environments and experiences themselves; and (4) the interaction techniques demanded by the operating systems and content.

Part 3 was a co-design exercise with the aim of understanding high-level user requirements in order to design more inclusive VR or AR experiences. This part involved a card sorting exercise in which participants were presented with a range of strategies that might help make immersive interfaces and content more usable and enjoyable for everyone. Participants were asked to reflect on which ideas they felt were the most important for them and then specify the two or three solutions that were most relevant to them individually. The frequency of these individual selections was used as the basis for subsequent discussion in the group to arrive at a final ranking of the most relevant solution.

Part 4 was an open discussion.

For the card sorting exercise, we asked participants to consider the following high-level accessibility strategies and choose the two most important to them. These were distilled from an online survey ($n = 101$) we had previously carried out and included the following ways to overcome access barriers:

- **Customisation:** Allows users to customise accessibility settings, for example, placement and styling of captions, scene contrast, and pacing of narratives.
- **Assistive Technology:** Allows users to benefit from their own assistive technologies, such as screen readers and switches.
- **Familiarisation:** Provides users with more tutorials and means of familiarisation, for example, offers greater assistance in learning about the technology and access to virtual environments that are familiar to users and thereby less confronting.
- **Adaptation:** Automatically adapts features to users' abilities, for example, by letting the content or system adjust itself, such as by adjusting the difficulty of a game or by making objects easier to reach, or more visible.

- **Awareness:** Allows users to maintain better awareness of the physical environment in which they engage with the immersive experience, for example, by making users aware of the physical world while not distracting them from virtual content.
- **Information:** Makes more information available to users, for example, provides hints and warnings, as well as presents the same information in multiple forms, such as captions or vibrations.
- **Interaction:** Supports more ways to interact, for example, by allowing users to select different modes of physical interaction, such as enabling interaction via hand tracking, head movement, controllers, etc.

3 Results

In this section we report on the results of the card sorting exercise as this was the primary focus of the workshop, and also generated the mostly lively and considered discussion among focus group participants. We group-elicited requirements by the common strategies listed above.

3.1 Virtual Reality

The most popular solution strategies among the focus groups (FGs) are Customisation (FG1, FG3, FG4) and Interaction (FG1, FG2, FG3). The next most prioritised is Information (FG2, FG4). In addition, one focus group also prioritises Adaptation (FG2).

Customisation: A priority area that emerged from the focus groups is enabling fine-grained adjustments of the visual appearance of the virtual world and any virtual controls, including brightness, contrast, inverted colour schemes, and other overall visual aspects (FG1, FG3, FG4). This area also includes the ability to adjust any text in terms of size and colour (FG1, FG3, FG4). Related, another highly preferred requirement is support for varying the speed of any animations, including scrolling text and video streams (FG1, FG3, FG4).

Other prioritised elements involve support for captions, subtitles, and audio descriptions of settings and surroundings (FG1, FG4). There was a desire for automatic transcripts (FG1), ideally coupled with the option to mix text and British Sign Language at various points (FG4).

One focus group (FG1) suggestion is to fine-tune interactions by allowing the user to control the speed of the pointer and click in the visual scene to access different areas (as opposed to having to navigate to reach them). Another suggestion is to include an option for an audio or vibration trigger to signal a request for user engagement.

One focus group (FG3) request is an ability to manually configure support for users with variable dexterity by, for example, allowing users to adjust the amount of movement required to reach virtual objects, in order to reduce fatigue.

In addition, we observe a range of other elicited user requirements: however, they are not as consistently prioritised across all focus groups.

Interaction: There is a desire to support multiple means of interacting. Eye gaze is identified as one such modality (FG1, FG3), as one switch-based accessibility systems, such as an eyebrow switch and gaming controllers (FG3).

Another prioritised group of requirements is voice control support, such as permitting users to use voice commands for positioning within VR and for navigating menus and interacting in general (FG2, FG3).

In terms of mixing and matching modalities, one focus group (FG2) explicitly suggests support for varying modalities according to current conditions, providing users with choices, but taking care not to force users to choose every single time the conditions change (FG2).

The remaining elicited requirements are largely in the area of ergonomics. We identify an area of high-level requirements relating to the ergonomics of the headset, such as supporting shoulder braces for the headset to assist the user in holding the head up and ease, or eliminate, neck strain (FG1). A related requirement is to reduce the weight of the headset (FG1) and to eliminate the need to strap the headset to the head (FG1). Headsets should also be untethered and thus not require wired connections (FG1). Last, ideally headsets would either provide an ergonomic experience for users needing to wear glasses or eliminate the need to wear glasses with the headsets (FG3).

In terms of hand control, headsets should work for users wearing gloves (FG1) and reduce neck and body movement by further hand controls, including gesture control (FG2, FG3). In terms of physical controllers, it is desirable to adopt the inclusive design approach and consequently implement light modifications to existing controllers to make them more accessible, as this may be more affordable for users than having to purchase expensive bespoke solutions (FG3).

Finally, ideally there should be no requirement to install sensors in a room, and if there is such a requirement, the systems should make it easier to instal such sensors (FG1).

Information: There is a desire to receive help spoken with a clear, pleasant voice and to avoid jargon (FG2). If it is necessary to introduce new concepts, then these should be clearly explained (FG2). Systems should provide a written and spoken tutorial on how to use them, suitable for first-time users (FG4). Another suggestion is to provide tutorials for each individual skill or task (FG2).

To encourage exploration, systems should have ways of allowing users to explore different options and settings (FG2).

Finally, systems should provide a way for users to assess their motion sickness in VR (FG2).

Adaptation: One focus group (FG2) reflects on the solution focusing on adaptation. One important aspect of such adaptation is enabling user control by providing mechanisms that allow the user to regulate the level of automatic adaptation and turn it off. In conjunction, it is suggested that it may be useful to have the option for users to provide data of their interactions and behaviours to improve system adaptation. Finally, users should be prompted when automatic adaptation may be useful and they should be allowed enough time to absorb new content and instructions before adaptation proceeds.

In terms of system-side requirements on adaptation there is a suggestion of supporting variable automatic adaptation for fluctuating conditions and a desire for the adaptation system to be able to ignore unintended movements and actions with high accuracy.

3.2 Augmented Reality

The most popular is Customisation (FG5, FG6, FG7). The second most prioritised solution strategy is Interaction (FG5, FG7). Finally, one focus group (FG8) prioritises Awareness.

Customisation: Similarly to VR, there is a strong desire to be able to fine-tune the visual appearance of graphical elements, such as adjusting the size and colour of all text (FG5, FG6, FG7) and changing colour, transparency, and contrast on all virtual content (FG5, FG7). Again, similarly to VR, there is a desire to vary the speed of animations and any scrolling text (FG6, FG7).

In terms of user interface organisation, there should be options to reduce choices and simplify the interface (FG5, FG6) and change the layout of virtual objects and controls (FG7). There should be an option to prevent pop-ups (FG7) and an option for sticky menus that remain in view until they are explicitly dismissed by the user (FG7). Finally, users should be able to set the field-of-view (FG5, FG7).

Audio design is another prioritised high-level requirement area. It should be possible to turn on and off audio, verbal prompts, subtitles and captions (FG5, FG6). In addition, there is a desire for the ability to configure audio cues to represent different elements of the design (FG5). There should be audio descriptions of text (FG5), and ways of probing the environment and receiving audio descriptions, such as "What is in front of me?" (FG5, FG6).

There should be support for British Sign Language in addition to text and audio (FG6) and the ability to enable or disable vibration cues (FG5).

In terms of ergonomics, it would be beneficial to be able to adjust the perspective for a user sitting upright or in a wheelchair (FG5). Object rotation could be supported by button interaction or manual input of degree of rotation, in addition to standard rotation interaction that requires the user to twist their wrist or fingers (FG5).

Interaction: All focus groups (FG5–FG8) desire support for offline editing of AR content and enabling interaction without the need to hold a device.

Again, similar to VR, a range of modalities are requested for input, including voice activation and voice commands (FG5, FG7), gaze interaction (FG7), and support for wearables to allow for longer reach and to enable alternative means of input and output (FG5, FG7).

In terms of output, there is a need for audio and haptic feedback (FG5), voice output of menus (FG5), and context-informative audio cues when traversing menus (FG5). As with VR, there is also a suggestion of functionality that allows users to point at a particular location and receive an audio description (FG). In addition, there is also a request to provide soundscapes when moving around (FG5). Finally, systems should have audio-only and visual-only modes (FG7).

Regarding user interface design, there should be a simple means of resetting the device, such as shaking the phone (FG5), and a mechanism for partitioning the user interface into separate chunks that can be used in isolation (FG7).

Relating to ergonomics, systems should support multiple ways of holding or mounting a device (FG5) and a variety of methods for users to hold equipment and tools, such as styli, to assist with dexterity and fine movements (FG7).

Finally, one focus group (FG7) expresses a requirement for systems to provide easy means for users to remain aware of their physical surroundings, which we will elaborate on when discussing the requirements for the next solution strategy: supporting awareness of the physical world.

Awareness: One focus group (FG8) considers this strategy. Systems should support a mechanism that allows users an easy way to leave AR and return to an unobstructed view of the physical surroundings. Related, another requirement is an ability to regulate the amount of virtual content to prevent virtual content clutter from obstructing the physical surroundings. Last, systems should maintain a view of virtual objects even when the user is moving, such as when using a wheelchair.

Systems should be transparent on who can view the user's surroundings and what these observers will perceive. In addition, systems should be aware of pavements and other surfaces, and their level of degradation, to ensure they can give users sufficient guidance to prevent accidents.

4 Discussion

This work represents a preliminary exploration of inclusivity requirements for VR and AR immersive content consumption. Our findings complement other efforts seeking to establish user requirements in this space. W3C's XR Accessibility User Requirements (2021) is an excellent attempt to document specific user needs and requirements. Oculus, a major VR headset manufacturer and content developer, also now offers developer guidance on Designing Accessible VR (2022). We contribute to the emerging understanding in this space by capturing and summarising the voice of the user. This includes providing insight into the prioritisation of inclusivity requirements that are otherwise typically presented as if they all have equal importance to the user.

We now briefly reflect on several limitations of this work. First, our assignment of participants to focus groups based on their dominant capability loss was imperfect due to scheduling constraints. This limits our ability to directly relate specific requirements to particular access needs. Nevertheless, a benefit of partly mixed groups was that the discussion could focus on high-level solution strategies that were broadly effective for different capability loss types. We consider this high-level discussion an advantage given the nascent stage of the requirements process.

Another important limitation of this work is that participants had varying levels of exposure to VR and AR content. This limited prior experience may have reduced the specific insights participants were able to bring. In addition, it is difficult to separate usability issues associated with encountering a new and unfamiliar form of technology from those usability issues arising from a given capability loss.

Future work involves using the identified high-level requirements as a basis for further investigation, including follow-up focus groups, to validate findings and elaborate further on user requirements, and iterative research around concrete solutions to access barriers to immersive content consumption.

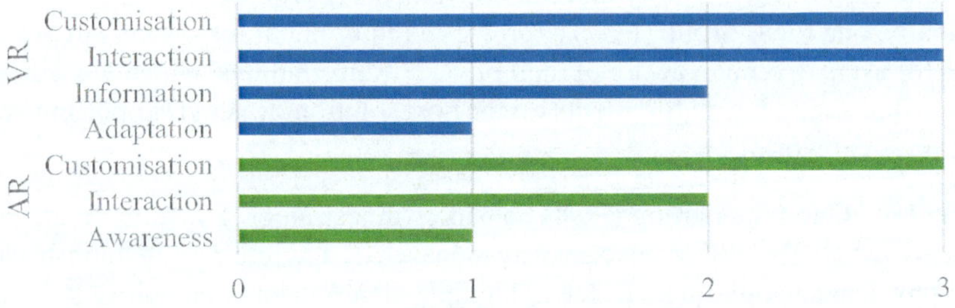

Fig. 1. The prioritised solution strategies for more inclusive VR and AR

5 Conclusion

We have reported the outcomes from eight focus groups involving participants with disabilities. We used card sorting to understand users' prioritised strategies for tackling barriers. Figure 1 summarises the number of focus groups listing each strategy as critical. For VR, we found that they were, in order of priority, customisation, interaction, information, and adaptation. For AR, we found that they were customisation, interaction, and awareness.

We clustered the elicited high-level requirements and found that customisation was consistently of the highest priority to users and providing this is critical for fully inclusive VR and AR content consumption.

We also found that interaction should be multimodal and allow users to mix and match hand tracking, controllers, gaze, and support for accessible technology, such as switch-based systems. In both VR and AR, users further desired means to point at various elements of an interface and be given descriptions.

In VR, information and help was another area identified as being of high importance and a range of high-level requirements emerged, such as providing initial guidance, offering spoken help with a clear voice with careful pacing, and ensuring there are tutorials in place explaining how to achieve specific tasks or goals.

In addition, in VR, one focus group identified adaptation as an important strategy that resulted in several high-level requirements, including the need to provide variable adaptation, means of turning it off, and means for users to provide deliberate training data to the system for adaptation. Finally, in AR, one focus group considered the strategy of allowing users to be aware of their physical environment to be important and suggested a range of requirements in this area.

These focus group results provide the basis for the prioritisation of subsequent efforts seeking to establish more specific requirements and corresponding technological solutions. We anticipate that this enhanced understanding of users' needs and wants within this design space will be of greatest benefit in the immediate term to developers of immersive content who currently lack effective guidance on making VR and AR content accessible.

Acknowledgements. This work was supported by the Engineering and Physical Sciences Research Council (Grants EP/S027432/1 and EP/S027637/1).

References

Clarkson, J., Coleman, R., Keates, S.: Inclusive Design: Design for the Whole Population. Springer, London (2003). https://doi.org/10.1007/978-1-4471-0001-0

Garaj, V., Dudley, J., Kristensson, P.O.: 5 ways the Metaverse could be revolutionary for people with disabilities. World Economic Forum (2022). www.weforum.org/agenda/2022/09/five-ways-the-metaverse-could-be-revolutionary-for-people-with-disabilities/. Accessed 16 Sept 2022

Garaj, V., Pokinko, T., Hemphill, C., Sesay, N.: Inclusive design of the immersive reality: eliciting user perspectives. In: Proceedings of the 7th International Conference for Universal Design (UD2019). International Association for Universal Design, Bangkok, Thailand (2019)

Keates, S., Clarkson, J.: Countering Design Exclusion: An Introduction to Inclusive Design. Springer, Heidelberg (2010)

Oculus: Designing accessible VR (2022). https://developer.oculus.com/resources/design-accessible-vr/2022-12-12. Accessed 12 Dec 2022

Ravenscraft, E.: What is the Metaverse, exactly? Wired (2022). www.wired.com/story/what-is-the-metaverse/. Accessed 24 Jan 2023

Shakespeare, T.: The social model of disability. Disabil. Stud. Reader **2**, 197–204 (2006)

Wong, A., Gillis, H., Peck, B.: VR accessibility: Survey for people with disabilities. Technical report (2017). https://drive.google.com/file/d/0B0VwTVwReMqLMFIzdzVVaVdaTFk/view?resourcekey=0-3X6VtbtDNvcbResutE_EVA. Accessed 9 Sept 2022

W3C: XR accessibility user requirements (2021). www.w3.org/TR/xaur/. Accessed 12 Dec 2022

XR Access: About XR access (2022). https://xraccess.org/about/. Accessed 24 Jan 2023

XR Association: Chapter three: Accessibility & inclusive design in immersive experiences (2022). https://xra.org/wp-content/uploads/2020/10/XRA_Developers-Guide_Chapter-3_Web_v3.pdf. Accessed 9 Sept 2022

Low-Carbon Behaviour in Older Chinese Migrants: Theoretical Models

Q. Ni[1(✉)], A. Kaniadakis[2], and H. Dong[1]

[1] College of Engineering, Design and Physical Sciences, Brunel Design School, Brunel University London, Uxbridge, UK
{qing.ni,hua.dong}@brunel.ac.uk

[2] College of Engineering, Design and Physical Sciences, Computer Sciences, Brunel University London, Uxbridge, UK
antonios.kaniadakis@brunel.ac.uk

Abstract. The existing low-carbon behaviour change theories have not paid much attention to the particularity of the cultural values and experiences of older migrants in the UK. This paper reviews the relationship and role of the Low-carbon Behaviour Model, Theory of Planned Behaviour, Resilience Model, Acculturation Model, Ecological Systems Theory, Behavioural Change Wheel, and other relevant theories in the low-carbon behavioural transformation of older Chinese migrants. An integrated model was proposed to explore net-zero to Chinese older migrants.

1 Introduction

All consumer activities that contribute to reducing an individual's CO_2 emissions can be considered low-carbon behaviours (Dubois et al. 2019). Older adults have played a leading role in driving greenhouse gas emissions. Between 2005 and 2015, their contribution to total national consumption emissions increased from 25.2% to 32.7% in 32 developed countries (Zheng et al. 2022). Despite the widespread attention to this issue, a definitive analysis of the challenges of balancing the needs of an ageing society with climate change mitigation is far from complete. Migration can also contribute to high carbon emissions. Most migrants come from developing countries or regions with lower per capita footprints; when they move to host countries with higher per capita footprints, migrants' carbon footprints may increase (Zheng et al. 2022).

Chinese migrants are a rapidly growing minority in the UK; they are one of the main minority groups (Gao et al. 2021). Traditional cultural values in China have an obvious role in promoting sustainable behaviour (Jiang et al. 2019). Context and environment influence and limit personal actions. When migrants have to adapt to different cultures, the language barriers they encounter, and the sense of well-being they experience all change with migration and the social environment of the place where they live (Suárez-Orozco 2012).

J. Goodman-Deane et al. (Eds.): CWUAAT 2023, *Design for Sustainable Inclusion*, pp. 22–30, 2023.
https://doi.org/10.1007/978-3-031-28528-8_3

Most studies of older Chinese migrants focus on their physical health problems (Gao et al. 2021), mental health problems, and life navigation problems (Caidi et al. 2020). There are few studies on their carbon emissions. As the first step to understand low carbon behaviours of older Chinese migrants, we reviewed a number of theoretical models and frameworks.

2 Methods

This research involves concepts such as older Chinese migrants, acculturation, and low-carbon behaviour. To better understand and establish the connection between them, we searched the library database of Brunel University. A total of 540 journal articles were found through searching "Older Chinese Migrant*" and "Elderly Chinese Migrant*", and 134 were related to theoretical models. In addition, a total of 214,000 journal articles were found by searching "Cultural Adapt*" and "Acculturate*", of which 38,200 were related to theoretical models. Besides, a total of 217,600 journal articles were found by searching "low-carbon", "Sustainable", "Pro-environment", and "Net-zero", of which 38,108 papers were related to theoretical models. After skimming through titles and assessing their relevance, and reading the selected abstracts, the first author chose around 20 models. Through discussion with the co-authors, a dozen models were selected for understanding the cultural adaptation process and behaviour changes of migrants. After looking into details of these models, the first author selected the final six as the most useful models to guide her PhD research.

3 Results

The selected models include the Low Carbon Behaviour Model, Theory of Planned Behaviour, Resilience Model, Acculturation Model, Ecological Systems Theory, and Behaviour Change Wheel. They are explained in the following three sections.

3.1 Low Carbon Behaviour Model and Theory of Planned Behaviour

Factors influencing low-carbon behaviour include interacting demographic characteristics, internal factors such as attitudes and habits, and external factors such as social norms and economics (Wang et al. 2021). The theory of planned behaviour (TPB) is an extension of the idea of rational action and is widely used as the basic framework in low-carbon behaviour studies. TPB as a framework for studying environmental behaviour can explain most intentions and future behaviour (Kaiser et al. 2005). According to TPB, personal attitudinal factors, normative factors, and perceived behavioural control determine one's intention to engage in a behaviour. Ajzen (1991) added behavioural, normative, and control beliefs as additional factors that influence these to form a new theoretical model.

The theory of planned behaviour (TPB) is an extension of the idea of rational action and is widely used as the basic frame in low-carbon behaviour studies. TPB as a framework for studying environmental behaviour can explain most intentions and future behaviour (Kaiser et al. 2005). According to TPB, personal attitudinal factors, normative factors, and perceived behavioural control determine their behavioural intention to engage in it. Ajzen (1991) added behavioural beliefs, normative beliefs, and control beliefs to form a new theoretical model (Fig. 2).

Research efforts have been made to improve the explanatory power of the theory. For example, the extended TPB model that includes moral obligations helps to explain Taiwanese energy-saving and carbon-reduction behavioural intentions to mitigate climate change (Chen 2016). Psychological factors are the focus of low-carbon behaviour. Cultural values as an antecedent variable of TPB have been validated in some cases (Wang and Wu 2016). Specifically, Chinese cultural values (e.g., man-to-nature orientation and collectivism) were added to the TPB model to better explain the home low-carbon consumption behavioural intention in China, and to look into the specific role of cultural values in the model. Anticipatory pride and guilt was also introduced to compensate for neglecting emotional factors (Jiang et al. 2020).

The above conceptual models explain what determines an individual's behaviour from different perspectives. The two theories (i.e., Low carbon behaviour model and the TPB) complement each other and have different emphases.

3.2 Resiliency Theory, Ecological Systems Theory and Acculturation Theory

Resilience is the process by which individuals use their abilities, strengths, and resources to bounce back by overcoming or actively adapting to various risky or adverse experiences that increase susceptibility to mental health symptoms. The components of resilience according to Mattelin et al. (2022) are courage, self-esteem, competence, self-confidence, capacity, trust, self-image, communication skills, and intelligence.

Ageing has been described as a time of resilience. Some negative factors of Resilience Theory are associated with migrants' backgrounds (Mattelin et al. 2022). For instance, the accumulation of factors such as limited English proficiency, withdrawal from the labor market, lack of social benefits, social isolation, cultural differences, and social status (Temple et al. 2022). These resilience-related negative factors appear decisive for older migrants. High resilience is significantly associated with positive outcomes, including successful ageing, lower depression, and a more active longevity (MacLeod et al. 2016).

Ecosystems describe multi-person and multi-system interactions between individuals and the environment at different levels (Bronfenbrenner 1992). For example, the microsystem includes interactions that affect the individual most directly, e.g., family, peers, and workplace. At increasingly distant levels, there are the mesosystem (interactions between microsystems), exosystem (e.g., social services, local politics, and mass media), macrosystem (attitudes and ideologies of the culture) and chronosystem (time). Ecological Systems relate to the interactions between migrants and different people and the macro-level and time-level factors that affect migrants (Tolentino et al. 2022). Migration will bring about changes in these. Different environments and cultural atmospheres may bring about changes in values for the older people (Kallis et al. 2022), such as changes in eating habits and changes in consumer habits. Changes in immigrant ecosystems will also affect carbon emissions in the lives of older migrants.

For migrants, cultural adaptation involves physical, psychological, cultural and social changes as they adapt (Liu 2019). The acculturation model (Ward et al. 2020) incorporates a wide range of variables. At the individual level are the characteristics of both the person (e.g., personality, experience, and values) and the situation/environment (e.g., cultural distance, and the amount and quality of intra- and inter-group contact). At the societal level are the variables of the society of origin and the society of settlement, including social, political, economic and cultural factors. All of these contribute to cross-cultural transition, stress, stress coping strategies, responses and outcomes.

3.3 The Behaviour Change Wheel

The Behaviour Change Wheel (Michie et al. 2011) is often used in the analysis and research of intervention design. It incorporates three levels. At the centre are the sources of behaviour (covering capability, opportunity and motivation). At the next level, there are intervention functions, such as education, incentivisation, and restrictions). At the outermost level lie the policy categories, such regulation, fiscal measures, and service provision.

With the emergence of behavioural change on low carbon politics and knowledge agendas, the question of how to understand and intervene in carbon-intensive practices becomes critical (Schwanen et al. 2012). Consumer behaviour is key to society's impact on the environment. Changes in social practices and behaviours will lead to a reduction in carbon-intensive lifestyles (Moloney et al. 2010).

4 Synthesis and Discussion

Low-carbon targets can be divided into different levels; they correspond to different goals which have various degrees of impact on social development and the living environment (Fig. 1). The high-level goals represent global sustainability, the middle-level goals represent the contributions that the community can make, and the low-level goals represent individuals, that is, those who have the long-term goal of net zero and will explore and practise low-carbon behaviours.

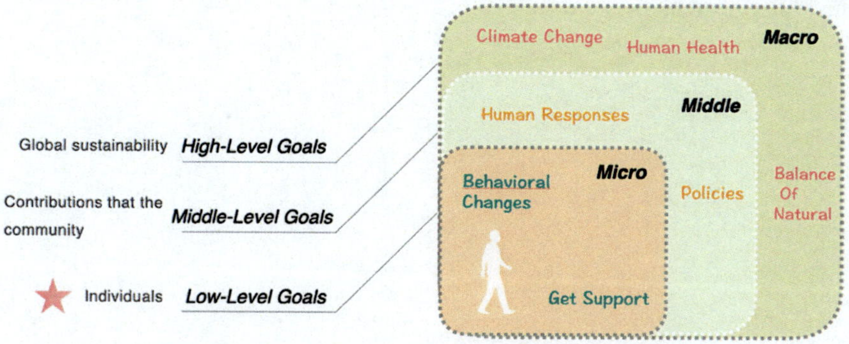

Fig. 1. Different levels of low-carbon behaviour goals

Most of the internal factors that affect low-carbon behaviour are also the influencing factors of TPB. The external factors that affect low-carbon behaviour change with society, and there are different variables in different situations. Situational factors (also known as external factors) influence variables that do not occur within an individual but come from elsewhere (Boldero 1995).

The influencing factors of Low Carbon Behaviour and Acculturation models have a lot in common. Therefore, they can be integrated into one for micro-level, individual behaviours. Ecological Systems Theory can provide a holistic perspective for other levels of action needs and the roles of relevant stakeholders. Figure 1 synthesises the factors influencing low carbon behaviour at different points in the system.

In TPB, the subjective norm refers to the impact of social pressure on individual intentions and behaviours. This is more important than other factors in China, which is based on the influence of traditional Chinese collectivist values and culture (Wang et al. 2014). Perceived behavioural control variables imply that factors such as opportunity, resources, time, knowledge, and skills are not under the control of the individual, thus influencing action intensively. Both morality and knowledge enable people to overcome barriers to engage in low-carbon behaviours (Chen 2016). Attitude refers to the understanding of behaviour results and the estimation of value. The traditional Chinese man-to-nature orientation value has an impact on attitudes.

Situational factors refer to external factors such as interpersonal influence (e.g., persuasion.), information, social norms, policies, economics, and technology. These external factors have mainly effects on individual energy consumption behaviour.

The diverse characteristics of the population also affect energy consumption behaviour (Echegaray and Hansstein 2017). Numerous studies have shown that age, income, living environment, and education level significantly impact on low-carbon behaviour.

The integrated model in Fig. 2 illustrates that low-carbon behaviour is affected by different factors at different levels (from the individual to the chronosystem); it provides a more comprehensive theoretical framework for studying low-carbon behaviour in older Chinese migrants.

Low-carbon research on older Chinese migrants should consider their changing cultural adaptability and the change in perceptions caused by their migration experience. A preliminary model (Fig. 2) was proposed to utilise relevant theories at different stages of the PhD research.

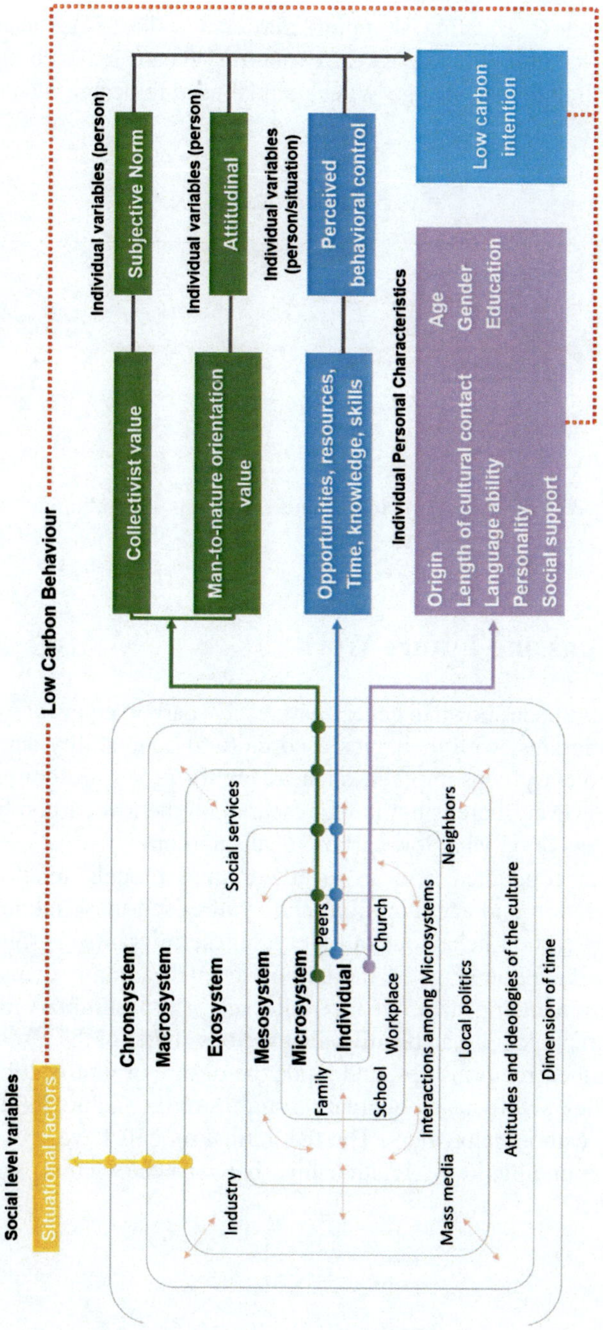

Fig. 2. An integrated model

In Fig. 3, Ecological Systems Theory 'oversees' the whole process. Individuals' resilience changes with the social and psychological changes caused by migration. Their consumption habits (e.g., eating, shopping) may change due to mental and environmental changes which Acculturation Theory can explain. TPB can predict changes in low-carbon behaviour. The Behaviour Change Wheel can be used to design interventions at a later stage.

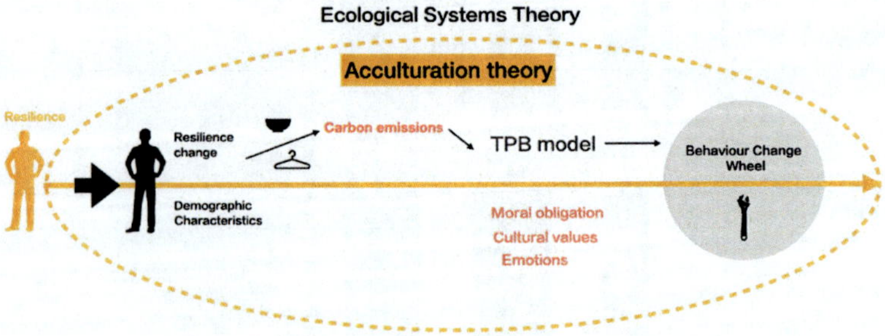

Fig. 3. Low-carbon to older Chinese migrants – a preliminary model

5 Conclusions and Future Work

Older people have been shown to generate excessive carbon emissions. There is evidence that Chinese migrants, with their increased cultural adaptability, are accompanied by higher-carbon dietary consumption, such as highly processed convenience food (Jin et al. 2017). However, there is not much research on the low carbon behaviour of older Chinese migrants, or on what low carbon means to them.

The authors conducted reviews of theoretical models relating to low-carbon behaviour, migration and acculturation, and synthesised these models to guide future research on low-carbon behaviour in older Chinese migrants. By integrating different models into a single one (Fig. 2), the authors created a comprehensive framework to understand the influencing factors at different levels. The preliminary model (Fig. 3) illustrates that different models can be utilised at various stages of the research to understand the factors and their relationships, and guide the design of interventions.

Future studies will focus on qualitative studies of the factors influencing older Chinese migrants' carbon behaviours. The theoretical models have provided the potential factors, and the findings (e.g., relationships between the factors) will help refine the integrated model.

References

Ajzen, I.: The theory of planned behaviour. Organ. Behav. Hum. Decis. Process. **50**(2), 179–211 (1991)

Boldero, J.: The prediction of household recycling of newspapers: the role of attitudes, intentions, and situational factors 1. J. Appl. Soc. Psychol. **25**(5), 440–462 (1995)

Bronfenbrenner, U.: Ecological Systems Theory. Jessica Kingsley Publishers, London (1992)

Caidi, N., Du, J.T., Li, L., Shen, J.M., Sun, Q.: Immigrating after 60: Information experiences of older Chinese migrants to Australia and Canada. Inf. Process. Manag. **57**(3), 102111 (2020)
Chen, M.-F.: Extending the theory of planned behaviour model to explain people's energy savings and carbon reduction behavioural intentions to mitigate climate change in Taiwan–moral obligation matters. J. Clean. Prod. **112**, 1746–1753 (2016)
Dubois, G., Sovacool, B., Aall, C., Nilsson, M., Barbier, C., et al.: It starts at home? Climate policies targeting household consumption and behavioural decisions are key to low-carbon futures. Energy Res. Soc. Sci. **52**, 144–158 (2019)
Echegaray, F., Hansstein, F.V.: Assessing the intention-behaviour gap in electronic waste recycling: the case of Brazil. J. Clean. Prod. **142**, 180–190 (2017)
Gao, S., Dupre, K., Bosman, C.: Understanding the neighbourhood environment and the health and wellbeing of older Chinese immigrants: a systematic literature review. Ageing Soc. **41**(4), 815–835 (2021)
Jiang, X., Ding, Z., Li, X., Sun, J., Jiang, Y., et al.: How cultural values and anticipated guilt matter in Chinese residents' intention of low carbon consuming behaviour. J. Clean. Prod. **246**, 119069 (2020)
Jiang, X., Ding, Z., Liu, R.: Can Chinese residential low-carbon consumption behaviour intention be better explained? The role of cultural values. Nat. Hazards **95**(1), 155–171 (2019)
Jin, K., Gullick, J., Neubeck, L., Koo, F., Ding, D.: Acculturation is associated with higher prevalence of cardiovascular disease risk-factors among Chinese immigrants in Australia: evidence from a large population-based cohort. Eur. J. Prev. Cardiol. **24**(18), 2000–2008 (2017)
Kaiser, F.G., Hübner, G., Bogner, F.X.: Contrasting the theory of planned behaviour with the value – belief – norm model in explaining conservation behaviour 1. J. Appl. Soc. Psychol. **35**(10), 2150–2170 (2005)
Kallis, G., Yarwood, R., Tyrrell, N.: Gender, spatiality and motherhood: Intergenerational change in Greek-Cypriot migrant families in the UK. Soc. Cult. Geogr. **23**(5), 697–714 (2022)
Liu, S.: English for international purposes: Guidance for Chinese students on an oversea sojourn. Rev. Educ. Theory **2**(4), 71–76 (2019)
MacLeod, S., Musich, S., Hawkins, K., Alsgaard, K., Wicker, E.R.: The impact of resilience among older adults. Geriatr. Nurs. **37**(4), 266–272 (2016)
Mattelin, E., Paidar, K., Söderlind, N., Fröberg, F., Korhonen, L.: A systematic review of studies on resilience and risk and protective factors for health among refugee children in Nordic countries. Eur. Child Adolesc. Psychiatry 1–34 (2022)
Michie, S., Van Stralen, M.M., West, R.: The behaviour change wheel: a new method for characterising and designing behaviour change interventions. Implement. Sci. **6**(1), 1–12 (2011)
Moloney, S., Horne, R.E., Fien, J.: Transitioning to low carbon communities – From behaviour change to systemic change: lessons from Australia. Energy Policy **38**(12), 7614–7623 (2010)
Schwanen, T., Banister, D., Anable, J.: Rethinking habits and their role in behaviour change: the case of low-carbon mobility. J. Transp. Geogr. **24**, 522–532 (2012)
Suárez-Orozco, C.: Identities under siege: immigration stress and social mirroring among the children of immigrants. In: Suarez-Orozco, C., Suarez-Orozco, M., Qin-Hilliard, D.B. (eds.) The new immigration, pp. 149–170. Routledge, NY, USA (2012)
Temple, J., Wilson, T., Brijnath, B., Utomo, A., McDonald, P.: English language proficiency among older migrants in Australia, 2016–2046. J. Int. Migr. Integr. **23**(2), 403–429 (2022)
Tolentino, J.-A., Jimenez, J., Conanan, A., Reano, C.: The Filipinx American ecological processes model: centralizing Filipinx values with an ecological systems approach. Asian Am. J. Psychol. **13**(1), 62 (2022)
Wang, T., Shen, B., Springer, C.H., Hou, J.: What prevents us from taking low-carbon actions? A comprehensive review of influencing factors affecting low-carbon behaviours. Energy Res. Soc. Sci. **71**, 101844 (2021)

Wang, J., Wu, L.: Research on household water conservation behaviour's response mechanism: extended model of theory of planned behaviour based on taoism perspective. Collect. Essays. Finance. Econ **207**, 105–112 (2016)
Wang, Z., Zhang, B., Li, G.: Determinants of energy-saving behavioural intention among residents in Beijing: Extending the theory of planned behaviour. J. Renew. Sustain. Energy **6**(5), 053127 (2014)
Ward, C., Bochner, S., Furnham, A.: The Psychology of Culture Shock. Routledge, Milton Park (2020)
Zheng, H., Long, Y., Wood, R., Moran, D., Zhang, Z., et al.: Ageing society in developed countries challenges carbon mitigation. Nat. Clim. Chang. **12**(3), 241–248 (2022)

Designing for an Ageing Population

Engagement Towards Creating New Models for Later Life Living

C. Zecca[1(✉)], C. McGinley[1], and R. Griffiths[2]

[1] The Helen Hamlyn Centre for Design, Royal College of Art, London, UK
cecilia.zecca@rca.ac.uk, chris.mcginley@network.rca.ac.uk
[2] Cartwright Pickard Architects, London, UK
r.griffiths@cartwrightpickard.com

Abstract. This paper presents initial findings and principles attained from an ongoing knowledge transfer project between academia and industry partners, aiming to develop more inclusive later-living housing models against the background of current UK market stagnation and lack of suitability of existing stock.

Housing for later-living in the UK lacks meaningful community engagement and hence frequently fails to embody genuine needs beyond basic accessibility principles. Potential residents are frequently reduced to simplified statistics or uncomplicated representations of 'third age'. The lack of engagement has contributed to unsuitable UK housing stock, inconsistent language use, and limited understanding of older peoples' actual experiences and desires for their homes and communities. The UK faces the challenge of creating more human-centric, socially and economically sustainable spaces within homes and in the urban environment, whilst avoiding age segregation. In recent years this has been acknowledged, yet developments claiming to be designed for older cohorts continue to be poorly executed, through fragmented planning policies; failed designs; persistently institutionalised features and lack of inclusive understanding.

This paper will discuss the prevalence of limited inclusive intentions and outcomes, particularly in terms of older populations, in architecture, and the consequences of neglected community engagement within the architecture design process, pointing out uncodified methodologies and fragmented literature. The paper will present a successful example of a housing project for later living in Europe, and conclude by proposing a more human-centric approach and a set of initial inclusive design approaches and principles for housing.

Keywords: Later-living · Housing · Human-centred architecture · Ageing

1 Introduction

Research indicates that there is a need to clarify the concept of inclusive design and associated guidelines within architecture and built environment industries (Ormerod 2005; Sugiyama and Ward Thompson 2007; Harding 2020). This lack of clarity on inclusive design process and methods in architecture has led to the marginalising of elderly groups

J. Goodman-Deane et al. (Eds.): CWUAAT 2023, *Design for Sustainable Inclusion*, pp. 33–41, 2023.
https://doi.org/10.1007/978-3-031-28528-8_4

through the routine development of institutionalised age-restricted housing complexes, so that segregation has doubled during the last 25 years (Kingman 2016).

Prevalent terminologies have shaped policy frameworks, generating confusion, inconsistency and limitations. The typical emphasis on accessibility can prove reductive in terms of expanding inclusive approaches in architecture practice (Harding 2020). Another example is the 'win-win' proclamations around on-trend 'solutions' such as 'downsizing', which in reality tend to maximise developers' return on investment, rather than genuinely respond to older people's aspirations.

Furthermore, despite generalised inclusion recommendations in UK housing policy, at a local level it is not clear what typologies are needed for new housing developments for older people (Stirling and Burgess 2021). There is a lack of knowledge about older people's experience of space both outside and inside their home, and a lack of clear process and principles towards designing inclusive buildings. While it is relatively common to find good examples of product and service design output achieved through inclusive processes, in architecture the typical 'Plan of Work' (RIBA 2020) has focused on technical design and performance of the building (Harding 2020) rather than inclusive process and output.

Based on literature, expert consultation, case study exploration and site visits, this paper proposes an inclusive implementation of RIBA's Plan of Work, with a focus on 'Stage 0 - Strategic Definition' through engaging architects with users. Further to this, three guiding principles are outlined for use in implementing Inclusive Design methods in Architecture. These principles highlight a range of spatial qualities and services that improve the human emotional, cognitive and multisensorial experience as a priority (Ritchie 2020), to be backed up by established and conventional intentions around space performance.

2 Literature: Representation in Architectural Inclusion

Although inclusive design has a long history in the UK context (Clarkson and Coleman 2015) discrepancies exist between academic recommendations and applied practices within architecture. Despite academic dialogue on user engagement and inclusive design strategies, within architectural practice there remains a distinct lack of understanding as to the benefits of such processes, and of approaches for applying them. In architecture practice, the narrative around participatory design processes is often dysfunctional, fragmented and self-referential (Jenkins and Forsyth 2010). Terms such as participation and co-design are underutilised, and conspicuous in their omission from the official RIBA (Royal Institute of British Architects) Plan of Work - the go-to process model within architectural practice, which include seven stages (0-Strategic Definition; 1-Preparation and Brief; 2-Concept Design; 3-Spatial Coordination; 4-Technical Design; 5-Manufacturing Construction; 6-Handover; and 7-Use). This is despite Inclusive Design being a recurring term highlighting good practice requirements by the RIBA (2020), in its architectural plan of work and stages: our research has identified that the knowledge and definition are often reduced to tokenism and simplification especially when concepts of less tangible disabilities, equity and differences are involved. There is a critical need to expand the scope of projects around housing to generate a more holistic definition of architectural inclusive design.

Compounding this issue is the media portrayal of older people in the UK, which often includes concepts of care, economic dependency and an unproductive population. This can negatively influence policy agendas and contribute to discrimination (Lloyd-Sherlock 2004). Schmid (2019) highlights a tendency to focus on numbers, statistics and potential housing options; however, the voice of older people is often devalued or totally absent: "*Those wishing to enter the sector that do not begin with a person-centred operations approach, instead skip straight to the numbers, largely fail to bring a product to the market*."

The only contact architects typically have with people and communities is their 'public consultation', which typically happens at a late stage in the design process, when planning permission, technical drawings (up to RIBA Stage 3) and related documentation have already been submitted to local planning authorities. Post Occupancy Evaluation (POE) is another form of engagement to collect users' input regarding buildings and spaces; however, as evidence demonstrates, it is still not fully integrated into typical design processes (Durosaiye et al. 2019).

Architectural participation and co-design require inclusive design approaches which, as previously discussed, are limited in the architecture disciplines (Zallio and Clarkson 2021). Inclusivity in its most simplified form is often equated to physical access for wheelchair users, neglecting other aspects of living in and experiencing space. On the contrary, inclusive methods, which are largely validated in other domains of design, encompass a variety of users with different needs, impairments and/or disabilities, and engage within multidisciplinary collaborations. The aim is to provide flexible and adaptable solutions that respond to the diversity of people across age, ability, gender and race. One key principle is to recognise exclusion through engaging with more extreme user experiences, understanding what bespoke solutions might be in those scenarios in order that they might be extended to all, acknowledging the diversity that occurs in reality (Zallio and Zanutto 2022).

While in service design, for instance, there is five steps methodology (Bhavnani and Sosa 2008) that enhances user engagement through observations and ethnographic approaches as well as a linear process to generate ideas, refine and implement design. In architecture this process has not been revised by RIBA and so not integrated in its Plan of Work.

In architectural practice there is a sense that the voice of less represented people is still not heard as it should be, and places and spaces are not actualising their potential, physically, emotionally or socially. This can be achieved by creating more intimate domestic spaces and healthier public spaces within the city, providing hybrid, sometimes deprogrammed spaces that can enable ownership and reappropriation of places for communities. In this scenario older people would be truly included in society, playing more active roles in the city and in the economy more generally. Architecture is not only about designing a building, but rather is about creating new relations and opportunities for people within the city where the building stands: "*Buildings should liberate their users, not limit them*". (Rogers 2017).

It is only by overcoming limited perspectives of integration and accessibility, of reduced standard dimensions and restrictions, and by looking at architecture through the magnified lens of inclusivity that we can truly improve people's lives by creating meaningful places and spaces.

3 Methodology: Literature Collation, Expert Consultation and Case Study Site Visits

To investigate the lack of inclusive design in architecture and the built environment, the research collated architectural articles and academic publications selected accordingly to: their originality in the ageing space; focus upon (predominantly) older participant engagement; and analysis of well conducted case studies of later living housing models. The review provided overview and insight into prior studies that specifically emphasize factors influencing the understanding of inclusive design in architecture for later living. The literature also provided an initial list of relevant case studies that were considered to have been produced with inclusive motivations, providing insight into the successful methods and perspectives relevant to business, architectural and built environment industries (Creswell 2003; Flyvbjerg 2006).

Next a theoretical framework for selecting the top-scoring case studies of those gathered was produced through three Delphi rounds, in order to validate quality-indicators (Hasson et al. 2000; Boulkedid et al. 2011) of best inclusive practice in architectural design for later living. The challenges, best practice and criteria for selection and assessment used to map and evaluate the case studies were refined through gathering the opinion of eight UK experts in health and social care, senior living and policies, affordable housing and sustaining places and inclusive design who were recruited and continue to actively advise on the project.

The wide collection of 60 international case studies enabled and supported the definition of a theoretical framework for applicable solutions, and exploration of potential adaption and adaptation in the UK sector. This is a validated method in architecture and built environment (Harding 2020).

The rational for case study selection was organised against nine criteria and their numerical evaluation through a yardstick appraisal (Yin 1994) as shown below:

The initial list of 60 potential case study sites was narrowed down to the 12 highest ranked examples that embodied current best practice in the sector.

Each short-listed case study was visited for fuller data collection, and to gather detailed accounts from those involved in the design, construction and management as well as residents in situ. In this regard, sets of audit criteria, questionnaires and semi-structured in-depth interviews related to inclusive design approaches, design challenges, residents' involvement and the financial and operational aspects of buildings were developed and later put through thematic analysis. The logistics of the site visits were organised through a 'site visit checklist' in order to standardise the process and enable analysis of the data collected by several researchers involved in the process.

The qualitative data obtained through these methods helped further develop a model with three main strands (Social, Spatial and Visual Connectivity; Sustainability and Diversity; Health, Wellbeing and Clarity). They were also translated into a briefing document, which acts as accompaniment to the RIBA Plan of Work.

Table 1. Criteria and evaluation for case studies selection

Criteria	Score	Weight
1. Location: Proximity to urban areas, amenities, GP, hospital, shops, services	0–5	10%
2. Design Solution: Volumes, internal spaces, light, outdoor spaces	0–5	15%
3. Inclusive Architecture: level of participatory design, residents' involvement since an early stage, use of workshops, focus groups	0–5	15%
4. Tenure & Mix of Uses: Innovative tenure model, affordability, BTR. Intergenerational, multipurpose spaces	0–5	15%
5. Technology: Assistive technology provision, social apps, motorised building elements	0–5	5%
6. Construction Methods: Modern method of construction (prefabrication, use of innovative materials	0–5	5%
7. Sustainable Solution: Use of low-impacting materials, energy efficiency (i.e. photovoltaic panels), rainwater collection	0–5	15%
8. Care Provision: Care provided on site, 24h extra care service, physical activities space	0–5	10%
9. Financial Model: Affordability, rental model, tenure mix, Government incentives	0–5	10%

4 Case Study Example: Agorhaverne Co-housing, Copenhagen, Denmark

The selection of best practices from worldwide housing developments was made against the criteria listed in the methodology section (Table 1), contributing to better understanding of process and elaboration upon the design principles. Beyond geographical location, the most successful later-living housing models presented recurring design themes and architectural inclusive design approaches as well as mixed functions and facilities. The example now presented demonstrates one exemplary case study from the 12 that provides a precedent to help develop a conceptual framework for pioneering housing models in the UK, as well as for improving the architectural RIBA work flow and stages. The case study presented is an affordable later-living development in Copenhagen, completed in 2021 and designed by Sangberg Architects.

Agorahaverne co-housing in Slagelse (Denmark) represents a successful inclusive design project where people, identity and nature complement each other and reinforce the sense of community within the development. One of the biggest challenges in terms of design, as architect Jonas Sangberg stated, was the interaction and balance between private and public spaces - "*the study of how they* [the residents] *pass through the space*". Compositionally (see Fig. 1), the layout is developed around an internal covered courtyard which works as a green house and central core of public life, with different spaces designed for social activities. The rhythm is articulated through a grid of timber columns, which reduce the scale of the open space usually and support the double pitched trusses and glazed roof. The central courtyard is also characterised by cubicle rooms, accessible from ground level but also inhabitable on the roofs, which can be used for meetings and social activities.

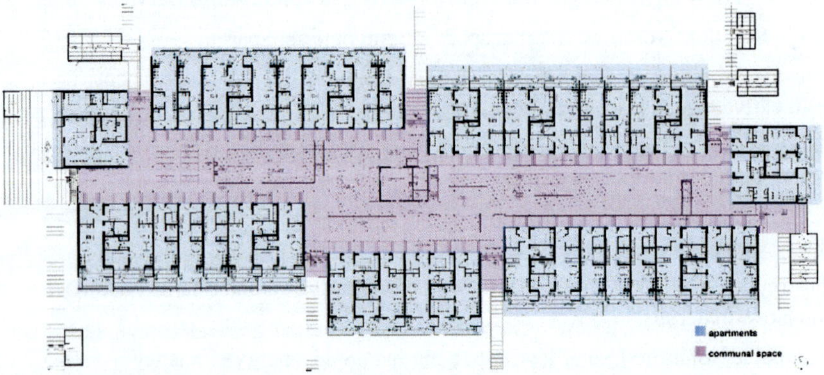

Fig. 1. Agorahaverne co-housing plan. Courtesy: Jonas Sangberg

Fig. 2. Agorahaverne courtyard view. Source: Author

The prefabricated apartments, all dual aspect, are distributed around the central void over two levels (see Fig. 2). In contrast to commonly found institutionalised aesthetic, the architect commented "*there has been important attention to detail and to internal materials in order to enhance the domestic feeling and experience*".

An innovative aspect in terms of architectural inclusive design is the collaboration between researchers, anthropologists and Urgent.Agency (a design company providing anthropological and demographical studies) which contributed to the brand of Agorahaverne through four main values: 1. Freedom, 2. Life-long learning, 3. Community, 4. Central square and public spaces.

The process of empathically engaging with potential and future residents was carried out by the developer Tetris together with anthropologists and researchers. This enabled analysis of the economy, the community aspirations and understanding of the dynamics of older people. During a site visit to Agorhaverne, this community-engaged approach was confirmed by residents - *"during the design stage we were asked what we like, why and how our days look."*

The level of engagement with local people also enabled future residents to know each other in advance and helped the developer to understand *"what we shouldn't do!"*.

5 Initial Findings

Findings from literature, case studies and thematic analysis of the interviews with the main stakeholders and residents led to the identification of two key aspects that need to be defined in the UK context: Architectural Inclusive Design Process and Architectural Inclusive Design Principles, still uncodified and underestimated in architectural practice for later-living housing models.

5.1 Architectural Inclusive Design Process

The research project identified the need for a profound shift in terms of age-inclusive approaches in Architecture. There needs to be understanding of older people, including the social context, physical surroundings and including new disciplines such as neuroarchitecture that highlight experience of space sensorially and emotionally (Mallgrave 2013). In terms of architectural projects, a deeper analysis, understanding site context and people's needs, is fundamental, and should be integrated in the RIBA Plan of work stage 0 - Strategic Definition. A multidisciplinary research approach undertaken from the beginning of the project (as we identified from the case studies and literature reviews) where architects engage with people and communities, can be effective in challenging unsuitable briefs towards better design intent. As architecture is for people, we conclude there are strong ethical grounds to position inclusiveness as a fundamental intent.

5.2 Architectural Inclusive Design Principles

The following working diagram proposes three initial areas of investigation and their interrelated subcategories that form guiding principles for a more inclusive and sustainable housing model for later-living. The main strands and their taxonomy of design principles summarise the findings collected and analysed so far from the case study projects (Fig. 3).

The first strand 'Social, Spatial and Visual Connectivity' helps to restore the sense of belonging to the place by creating more relation to the outdoor environment and community. This can be achieved in ways such as opening buildings to the public; increasing the mix of uses, and the provision of more accessible levels and comprehensiveness of care. To increase the connectivity between people, and between people and spaces, flexible open spaces able to adapt to different situations as well as various services should be included in early stages of the project. The second strand 'Sustainability and Diversity'

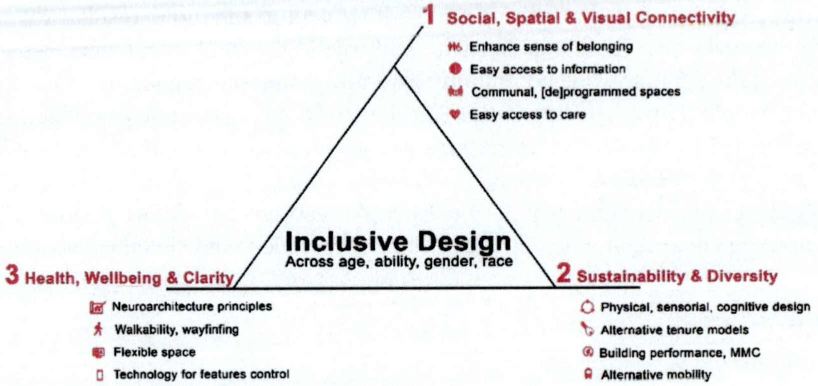

Fig. 3. Inclusive design principles for later living housing model. Source: Author.

is about connecting people within a more self-sufficient environment through design that considers the sensorial experiences of the spaces. Moreover, the model aims to offer more flexible and alternative tenure mixes by proposing alternative rental fees for the private spaces, facilities and semi-public spaces. The third and final strand 'Health, Wellbeing and Clarity' includes a more domestic design of clear wayfinding and communal areas that facilitate healthy balance between social engagement and personal life.

6 Conclusion

This paper discusses the lack of inclusive intent and community engagement during the initial stages of architectural design proposals for later living housing developments in the UK.

Some of the major findings are presented with the support of an exemplar case study provided as a highly ranked concrete example of our inclusion aspirations, which enables further discussion around the proposals for Inclusive Design Processes and Principles in architecture for later-living.

Specifically, this paper identifies, in terms of Architectural Inclusive Design Process, the need for codified forms of community engagement such as interviews, focus groups and studying users' journey during the initial stage of the design; the need for a more collaborative process and greater range of experts, such as researchers, within the development team.

In terms of Architectural Inclusive Design Principles for later living housing models, findings suggest three main areas (Social, Spatial Connectivity; Sustainability; Health and Wellbeing) where architects and other stakeholders can focus during the design process. These three areas are also elaborated upon through four subcategories of actions that can significantly enhance the design of a more inclusive housing model for later living.

Acknowledgements. This KTP (Knowledge Transfer Partnership) project is funded by Innovate UK. We are grateful to stakeholders and residents of Agorahaverne co-housing in Copenhagen.

Specifically: Betina Nørby Steinbock (Tetris A/S); Jonas Sangberg (Sangberg Architects); Ricky Storm Braskov (Urgent.Agency); Tomas Jandorf (Tetris A/S).

References

Bhavnani, R., Sosa, M.: IDEO: Service Design. Insead, Paris (2008)

Boulkedid, R., Abdoul, H., Loustau, M., Sibony, O., Alberti, C.: Using and reporting the Delphi method for selecting healthcare quality indicators: a systematic review. PLoS ONE **6**(6), 1–9 (2011). https://doi.org/10.1371/journal.pone.0020476

Clarkson, P.J., Coleman, R.: History of inclusive design in the UK. Appl. Ergon. **46**(Part B), 235–247 (2015)

Creswell, J.W.: Research Design: Qualitative, Quantitative and Mixed Methods Approaches. Sage Publications, Thousand Oak (2003)

Durosaiye, I.O., Hadjri, K., Liyanage, C.L.: A critique of post-occupancy evaluation in the UK. J. Housing Built Environ. **34**(1), 345–352 (2019). https://doi.org/10.1007/s10901-019-09646-2

Flyvbjerg, B.: Five misunderstandings about case-study research. Qual. Inq. **12**, 219–245 (2006)

Harding, J.: Moving inclusively through transport buildings: a cross-disciplinary design case study. In: Langdon, P., Lazar, J., Heylighen, A., Dong, H. (eds.) CWUAAT 2020, pp. 115–122. Springer, Cham (2020). https://doi.org/10.1007/978-3-030-43865-4_12

Hasson, F., Keeney, S., McKenna, H.: Research guidelines for the Delphi survey technique. J. Adv. Nurs. **32**(4), 1008–1015 (2000)

Jenkins, P., Forsyth, L.: Architecture, Participation and Society. Routledge, Milton Park (2010)

Kingman, D.: Generations apart? The growth of age segregation in England and Wales (2016). www.if.org.uk/wp-content/uploads/2016/09/Generations-Apart_Report_Final_Web-Version-1.pdf. Accessed 18 Jan 2023

Lloyd-Sherlock, P.: Living Longer: Development and Social Protection. UNRISD. Zed Books, London & New York (2004)

Mallgrave, H.F.: Architecture and Embodiment. The Implications of the New Sciences and Humanities for Design. Routledge, Taylor and Francis, Milton Park (2013)

Ormerod, M.: Undertaking access audits and appraisals: an inclusive design approach. J Build Appraisal **1**, 140–152 (2005)

RIBA: RIBA plan of work (2020). www.architecture.com/knowledge-and-resources/resources-landing-page/riba-plan-of-work. Accessed 18 Jan 2023

Ritchie, I.: Neuroarchitecture. Design with Mind in Mind. Architectural Design. Wiley, Hoboken (2020)

Rogers, R.: A Place for All People. Canongate Books Limited, Great Britain (2017)

Schmid, P.: The market needs greater choice, greater options. In: Just living. Collado Collins Architects (2019). www.colladocollins.com/publications/2019/just-living. Accessed 18 Jan 2023

Stirling, P., Burgess, G.: Understanding supply, demand and investment in the market for retirement housing communities in England. Cambridge Centre for Housing & Planning Research, University of Cambridge, UK (2021)

Yin, R.K.: Case Study Research: Design and Methods. Sage, Beverly Hills (1994)

Zallio, M., Clarkson, P.J.: On inclusion, diversity, equity and accessibility in civil engineering and architectural design. A review of assessment tools. In: Proceedings of the International Conference on Engineering Design (ICED 2021), Gothenburg, Sweden, 16–20 August 2021 (2021)

Zallio, M., Zanutto, O.: Housing a misura di senior. Progettare abitazioni age-friendly. Maggioli Editore (2022)

Promoting Physical Activity to Older Adults Using Different Ageing Models

J. Guan[1(✉)], E. Zitkus[1], C. de Bont[1], and H. Dong[2]

[1] School of Design and Creative Arts, Loughborough University, Loughborough, UK
{j.guan4,e.zitkus,c.j.de-bont}@lboro.ac.uk
[2] College of Engineering, Design and Physical Sciences, Brunel Design School, Brunel University London, Uxbridge, UK
hua.dong@brunel.ac.uk

Abstract. Promoting physical activity (PA) to older adults is beneficial for healthy ageing and addresses challenges posed by population ageing. The variable characteristics of older adults makes it difficult for stakeholders to understand and address their needs. This study aims to find out the advantages and disadvantages of current ageing models and ways to promote PA for all populations. Literature related to ageing models and PA was reviewed, and 48 older adults from different places in the UK were interviewed. The challenge of inclusivity in ageing and four common ageing models were discussed. Finally, the study presents four ageing models to propose ways of promoting PA for people. By increasing the inclusivity of ordinary communities, the naturally occurring retirement community (NORC) could be appropriate for healthy ageing. Future researchers may want to do the data collection in a more systematic way to compare all the ageing models.

Keywords: Physical activity · Healthy ageing · Ageing models · Inclusivity

1 Using Physical Activity to Address Inclusivity in Ageing

The rapid growth of the elderly population has become a global challenge. Population ageing challenges the healthcare industry as well as presenting opportunities for business expansion. For example, the emergence of healthcare industries has brought economic benefits to society and contributed to healthy ageing at the same time.

Healthy ageing, defined as "*The process of developing and maintaining the functional ability that enables wellbeing in older age*", has become the focus of the World Health Organization (WHO 2020). It is important to ensure that older adults are capable of living independently.

Older adults' diversity is reflected in the different needs they may have. The Ministry of Civil Affairs of the People's Republic of China provided a classification method according to older adults' health status. They can be classified into three levels: self-cared elderly, device-aided elderly, and nursing-cared elderly (MCA 2001). The self-cared elderlies are fully capable and can take care of themselves. The device-aided

J. Goodman-Deane et al. (Eds.): CWUAAT 2023, *Design for Sustainable Inclusion*, pp. 42–49, 2023.
https://doi.org/10.1007/978-3-031-28528-8_5

elderlies need handrails, crutches, wheelchairs, lifts, etc. for daily living. The nursing-cared elderlies have to rely on others for daily living, and they usually need medical care. Although older adults can be classified according to their physical capability, it is difficult to provide a one-size-fits-all solution for them all because of the great variability of older adults.

So there are other factors to be considered to implement inclusivity in our societies, which are outlined below: 1) Cognitive Capability: there is no typical older person (WHO 2021). At the same time, this classification is dynamic. When considering improving inclusion for older adults, physical and mental needs are equally important. However, compared with physical requirements, mental and emotional needs of older adults are more likely to be ignored. 2) Socio-Economic Differences: health inequalities arc widespread. Socio-Economic Status (SES) is a combined measurement of a person's education, income, occupation, and other factors. People of high SES have more resources than people of low SES, such as better living conditions and medical resources (Stalsberg and Pedersen 2018). However, it does not mean that people with low SES cannot achieve healthy ageing. Making society more inclusive will help to reduce inequality. It is also worth thinking about how to make the best possible services for older adults based on their current economic conditions.

Considering that more older people will have to share fewer resources in some developing countries such as China, more societies will need to think strategically to be inclusive. For example, the data from the National Health Commission of the People's Republic of China (NHC) shows that in 2021 there were 190.64 million people aged 65 and above, accounting for 13.50% of the total population (NHC 2021). Based on the demographic structure, the NHC proposed the "90-7-3 pattern", i.e., 90% of older adults will age in place, 7% of older adults will age in communities, and the last 3% will age in nursing homes. However, the government can only provide limited support for people ageing in place because of dispersion and cultural differences. In this case, physical activity can be a strategy to follow to enhance longer and healthier life.

Physical activity (PA), as one of the most effective, low cost, and sustainable solutions should be taken seriously. Among the many factors impacting ageing, studies have shown a positive connection between PA and healthy ageing (Daskalopoulou et al. 2017). A body of research has indicated that maintaining a reasonable level of PA weekly is beneficial for the health for all age groups (Davies et al. 2019) and it needs to be promoted everywhere for all ages and abilities.

2 Four Ageing Models

There are four common ways of ageing depending on the living environment: 1) ageing in place; 2) naturally occurring retirement community, 3) retirement community, and 4) care home. No matter where and how you age, remaining physically active is essential to keep healthy and independent.

Ageing in place can be understood as "remaining living in the community chosen by the person, with some level of independence rather than in residential care" (Horner and Boldy 2008; Wiles et al. 2012). In other words, ageing in place is ageing in the person's own home or property. The benefits of ageing in place include maintaining independence, autonomy, and social connections (Wiles et al. 2012). More importantly, ageing in place is the most affordable and inclusive way for the majority. The key element of ageing in place is that older adults should be able to keep their independence, which will not always be the case. The benefits of ageing in place should not be an excuse for government not to build specific care facilities. Older adults will need aid and care as time passes.

A retirement community, usually known as a continuing care retirement community (CCRC), provides service and resources for all self-care, device-aided, and nursing-cared older adults. The idea of CCRC was originally proposed in the United States. The first CCRC, Sun City, was constructed by Del Webb in Arizona State, United States in 1960. They stipulated that all residents must be over 55 years old, and the rule applies still today. They were promoting the idea that retirement should be start of "golden years" instead of waiting for death. After half a century of development, CCRC numbered approximately 1900 in the United States in 2010 (Shinan-Altman et al. 2020).

As opposed to CCRC which is specifically built for older adults, the naturally occurring retirement community (NORC) was not originally designed for older adults but eventually attracted a lot of them (Jiaxuan et al. 2022). The concept of NORC was originated in the 1980s, in Georgia, United States (Ivery and Akstein-Kahan 2010; Jiaxuan et al. 2022). Since the NORC is not organised by company or government, studies are still trying to clarify the definition. It is commonly agreed that to become a NORC, more than 50% of residents should be over 60 years old (Hunt and Ross 1990; Yalowitz and Bassuk 1998).

In order to provide optimal experiences for older adults for them to live independently and comfortably, the United States proposed the NORC supportive service programs (NORC-SSPs) in New York City in 1986 (Jiaxuan et al. 2022). To make sure that the NORC-SSPs are effective, there are four objectives proposed by New York State (Maclaren et al. 2007). First, the services should be effective and integrated into the community in order to meet the diverse need of residents. Second, preventive care and services are necessary to keep residents independent; they can also help to reduce long-term care which is more expensive. Third, the residents and social workers should be encouraged to participate in the decision process. Finally, the services should be personalised according to the residents' characteristics by involving them in the decision-making process such as by organising focus groups and co-design workshops. Although the NORC is not under the jurisdiction of the government or construction company, it will require funding to keep the NORC-SSPs on going, and the money usually comes from the government, based on the experience of New York.

When older adults lose the ability to live independently, they will need to be taken care of by family or care workers either at home or in a care facility. According to a report done by Laing (2017, 2018), a business intelligence provider across health, care and education, there were around 421,000 people aged 65 or above living in care facilities in the UK, with 40% of those self-funding. Additionally, 650,000 people in the UK were receiving homecare services, which excluding people who received NHS funded care service at home. According to the Care standards Act (UK Parliament 2000), care were defined as "provid[ing] accommodation, together with nursing or personal care, for persons who are or have been ill, who have or have had a mental disorder, who are disabled or infirm, or are or have been dependent on alcohol or drugs".

It is essential for people who are less capable when getting old or suffering from certain diseases, such as dementia, to live in care facilities being taken care of both physically and mentally. However, barriers such as lack of adequate communication, being physically inactive, limited space, and fear of falling still exist and those will make residents frailer (Chen 2010; Kalinowski et al. 2012; Benjamin et al. 2014).

For older adults, there is no best way of ageing, only the most suitable way for them. And it will depend on their physical capability, economic conditions, personal characteristics, family states, and their own lifestyles. It is also worth mentioning that the situation keeps changing as people get older. When an older person undergoes a major change, such as losing their partner or suffering from disease, they will need to re-evaluate their status and possibilities and sometimes choose another way of ageing. Different strategies should be applied to promote PA for older adults using different ageing models.

3 The Study: Data Collection and Analysis

This study was a literature review-based study and empirical data from interviews were used to test part of the results. This is because the study is part of an ongoing PhD research which is expected to finish in 2023. Literature related to ways of ageing based on living environments were reviewed, then four ageing models were explained. Two of the ageing models, ageing in place and CCRC, were focused on and discussed.

In order to find out how to promote physical activity (PA) for older adults using different ageing models, semi-structured interviews were conducted during an 8-months period. The structure of the interviews was based on the factors related to PA adoption including physical capability, social participation, technology, living environment, motivation, and enjoyment. A total of 48 participants were recruited including 27 people ageing in place from Loughborough, Leicester, and London, 15 people living in two CCRCs, Gifford Lea and Charles Waterton House, and 6 volunteers or personal trainers.

The interview was conducted both online via Zoom and Microsoft Teams and in-person depending on the preference of participants. Most interviews were around 30 min, some took longer if the person was talkative. All participants were provided with participants' information sheet and consent form which was reviewed and approved by Loughborough University's Ethics Review Sub-committee. All the sessions were audio recorded and transcribed, and NVivo was used for data analysis.

There were two factors that most participants emphasised during the interviews, 1) the social aspect of the activity; 2) the accessibility and inclusivity of the activity. The social side of activity is essential. In fact, some participants consider being socially active is more important than being physically active. They would prefer to join a group with other participants of a similar age so that they can have a common topic to talk about. However, they will not refuse to let younger ones to join their circle as long as they are not the only few seniors. It is worth mentioning that more than half of the older adults in this study were involved in voluntary activities such as walking group leader, research group participant, committee member of charity etc., which is related to social needs and self-actualisation. Accessibility and inclusivity are related to the location, time, expense, and eligibility of the activity or event. It was common for the participants to give up an activity over a long commute and time conflict. The expense was also one of their primary considerations since most of them were pensioners, and they had to reflect on their financial status because of increasing inflation. Some participants also mentioned that they could not get a membership from the local leisure centre because of their medical condition. The leisure centre or other local activities are mainly designed for the general public. People who are less capable or suffer from certain illness may find them not inclusive enough. The situation was much better in the CCRCs because they have well-being navigators who can provide one-to-one support for people with needs.

4 Discussion

Ageing in place is the most common ageing model. Older adults live all over the city and they should be able to access all resources including public parks, gyms, leisure centres and clubs. However, it is not easy to design a specific intervention for them since they live all over the city and their capabilities are diverse. Thus, interventions for people ageing in place can only be general and less effective. For example a leisure centre is built for general public not for people who have disabilities or suffer from certain diseases A leisure centre will need a medical certificate if they think the customer is too fragile to use the facility, which stopped one of the interviewees being physically active. Another participant admitted that income prevents them having a membership at a local leisure centre. It is the government's responsibility to make infrastructures inclusive enough for the public and consider how to encourage older citizens to be more physically active. For example, making the city more age-friendly can encourage older adults to engage in society since PA is not only about exercise but also related to activities of daily living (WHO 2007). More policy support such as bus passes, i.e., people can travel by bus for free if they reach 60 or the state pension age could be proposed, and it is also helpful to involve other stakeholders like charities and non-profit organisations.

Older adults can maintain their autonomy and independence when living in CCRC. The first author visited Gifford Lea, a CCRC of Inspired Villages Group Ltd, which has a well-equipped gym with two fulltime wellbeing navigators, a relaxation swimming pool, and an outdoor area for exercise. They invited fishers, barbers, and cosmetologists to come to the community regularly for their residents. Most daily needs can be fulfilled while living in this community, which is what the company and the residents are looking for.

However, it is easy to form a relatively closed environment, especially when all the residents are older adults. The CCRC can prevent some older adults, especially those who have mobility issues, from engaging with society. According to the interviews conducted with older adults living in CCRC, it can be difficult for them to visit their family or go shopping if they rely on public transport. A resident in CCRC mentioned that: *"Now I can drive so I can visit Chester or Liverpool whenever I want, like the jazz band I mentioned. I cannot imagine what I am going to do without my car. Although we have a bus service here, it is not convenient and [it is] unpredictable. The transportation service from the community must prioritise those with medical needs."* Not everyone living in that community can drive, and some of them have not left the community for long time. Additionally, the CCRC can become dull if there are no more new people moving in, since current residents inevitable become less capable.

Since older adults living in the CCRCs need to pay a service fee, funding is not the primary concern when organising events. CCRCs usually have facilities such as pools and gyms and they will organise PA events regularly. However, it is always up to the residents if they want to use the services or resources or not. It is important for CCRC to find the right motivation for their residents and provide different levels of PA for residents to address their specific needs. Since CCRC are normally far away from the city and public transport is not convenient, they should consider building connections with other stakeholders such as the local council, charities, and universities to increase openness to the community, especially for those who are not able to engage with activities by themselves.

Compared with ageing in place, promoting PA for older adults living is NORC can be easier since they live in relatively concentrated areas, especially when the government provide NORC-SSPs for residents such as recreational and educational programs, volunteer opportunities, and exercise groups etc. At the same time, NORC seems to retain all the benefits from ageing in place since it still is a normal community in the city and the residents do not need to pay extra fees for all the services they receive. However, it is important to make sure that the NORC-SSPs are sustainable, for example, encourage self-organised and voluntary activities. Otherwise, the activities are at risk of stalling if the government cut the budget.

It is common that people living in a care facility lack motivation for doing PA since they do not see how it can be helpful. Family members and care workers need to find what each person values and come up with a way to persuade and encourage them. One of the interviewees mentioned: *"When my husband lived in the care home, he refused to do any exercise. But I know what he wanted is just to get back to the car again. So, I said to him that, you have to do the stretch, and turn your neck to recover, so that you can drive again. And he did that. But I know he probably cannot drive forever. But the exercise is good for him, you know, the stretches."*

Care facilities should deliver tailored training for people suffering from certain diseases, as well as rehabilitation. Even for people who have very limited mobility, such as wheelchair users, PA is still essential. Since their residents are considered a vulnerable group, they should consult physiotherapists regarding appropriate activities to maintain their functional capability and avoid further injuries.

5 Conclusion

The government could provide support for older adults ageing in place and living in NORC. There are benefits to encouraging older adults to participate in society while they are still independent and healthy. The CCRC have better environment and resources, but only a few people can afford to live there. It is worth thinking that the CCRC not only benefits its residents but also contributes to people in that area by sharing its facilities. The collaboration could increase the openness of CCRC as well, otherwise the community will be very quiet and less active since residents are getting older as time goes by. Charles Waterton House, a CCRC near Wakefield, is collaborating with the local council and a primary school nearby to organise events together and the residents are excited to get themselves involved. Care facilities need to be more cautions when promoting PA since they are dealing with the most vulnerable group: inappropriate plans might harm them both physically and mentally.

Against the background of population ageing, more efforts should be made to keep society working smoothly and properly. After considering the pros and cons of implementing PA in the four ageing models, the NORC model appears more inclusive and by its integration with local communities by being more affordable and potentially an effective way for government to promote healthy ageing. It is cheaper for people to move and live in NORC compared with moving to CCRC if they feel they should get more support or they cannot manage cope with ageing in place, and the benefits of ageing in place could be retained. In this way, by providing NORC-SSPs, the government could provide localised support to more people using less resources. NORC-SSPs standards could be achieved if there are enough of them and connections can be built for them to share resources. At the same time, the network can also benefit people ageing in place because they can easily access the services near them.

PA is not only about exercise: activities of daily living are also considered as PA. Older adults can participate more in society in an age-friendly city, for example by making new friends and doing voluntary activities. It is beneficial to reduce social stigma and promote active ageing, which can help to address population ageing issues.

It is worth mentioning that data collection should be conducted in a systematic way to compare all the ageing models. Currently, researchers usually focus on one ageing model, and it is difficult to compare the result when participants are from different backgrounds. This study only interviewed people ageing in place and living in CCRC. One of the reasons is that the concept of NORC is not well accepted in the UK: most studies were conducted in the US and very few researchers in the UK are doing research on NORC and NORC-SSPs. Some interviewees' neighbourhoods may meet NORC requirements; however, it was still difficult for the authors to classify them based on existing data. Researchers may consider developing a short survey to evaluate if older adults' living environments are considered as NORC. Local authorities such as Borough Councils, might want to be involved in this as well because they will be able to develop tailored NORC-SSPs if they have the data to support their proposal. Also, future researchers should bear in mind that getting in touch with older adults living in care homes will require going through a stricter ethics approval process. However, more convincing findings will be discovered if the study is planned in more systematically which can bring potential collaboration opportunities between the government, CCRCs, and care homes, and NORC development in the UK.

References

Benjamin, K., Edwards, N., Ploeg, J., Legault, F.: Barriers to physical activity and restorative care for residents in long-term care: a review of the literature. J. Aging Phys. Act. **22**(1), 154–165 (2014)

Chen, Y.: Perceived barriers to physical activity among older adults residing in long-term care institutions. J. Clin. Nurs. **19**(3–4), 432–439 (2010)

Daskalopoulou, C., Stubbs, B., Kralj, C., Koukounari, A., Prince, M., et al.: Physical activity and healthy ageing: a systematic review and meta-analysis of longitudinal cohort studies. Ageing Res. Rev. **38**, 6–17 (2017)

Davies, S.C., Atherton, F., McBride, M., Calderwood, C.: UK Chief Medical Officers' physical activity guidelines. Department of Health and Social Care, pp. 1–65 (2019)

Horner, B., Boldy, D.P.: The benefit and burden of "ageing-in-place" in an aged care community. Aust Health Rev. **32**(2), 356–365 (2008)

Hunt, M.E., Ross, L.E.: Naturally occurring retirement communities: a multiattribute examination of desirability factors. Gerontologist **30**(5), 667–674 (1990)

Ivery, J.M., Akstein-Kahan, D.: The naturally occurring retirement community (NORC) initiative in Georgia: developing and managing collaborative partnerships to support older adults. Adm. Soc. Work **34**(4), 329–343 (2010)

Jiaxuan, E., Xia, B., Suslawati, C., Chen, Q.: An overview of naturally occurring retirement communities (NORCs) for ageing in place. Buildings **12**(5), 519 (2022)

Kalinowski, S., Wulff, I., Kölzsch, M., Kopke, K., Kreutz, R., et al.: Physical activity in nursing homes—Barriers and facilitators: A cross-sectional study. J. Aging Phys. Act. **20**(4), 421–441 (2012)

Laing, W.: Care homes for older people market analysis and projections. LaingBuisson, London, UK (2017)

Laing, W.: Care homes for older people: UK market report. LaingBuisson, London, UK (2018)

Maclaren, C., Landsberg, G., Schwartz, H.: History, accomplishments, issues and prospects of supportive service programs in naturally occurring retirement communities in New York state: lessons learned. Hous. Elderly: Policy Pract. Issue **4372**, 127–144 (2007)

MCA: Standards of social welfare institution for the elderly, MZ008-2001. The Ministry of Civil Affairs of the People's Republic of China (2001). www.elinklaw.com/zsglmobile/lawView.aspx?id=74445. Accessed 18 Jan 2023

NHC: 2020 National Aging Development Bulletin (2021). www.nhc.gov.cn/lljks/pqt/202110/c794a6b1a2084964a7ef45f69bef5423.shtml. Accessed 21 Aug 2022

Shinan-Altman, S., Gum, A.M., Ayalon, L.: Moving to a continuing care retirement community or staying in the community? A comparison between American and Israeli older adults. J. Appl. Gerontol. **39**(11), 1221–1229 (2020)

Stalsberg, R., Pedersen, A.V.: Are differences in physical activity across socioeconomic groups associated with choice of physical activity variables to report? Int. J. Environ. Res. Public Health **15**(5), 992 (2018)

UK Parliament: Care Standards Act 2000. The Stationery Office, London, UK (2000)

WHO: Global age-friendly cities: A guide. Community Health, p. 77 (2007). https://apps.who.int/iris/bitstream/handle/10665/43755/9789241547307_eng.pdf?sequence=1&isAllowed=y. Accessed 18 Jan 2023

WHO: Ageing: Healthy ageing and functional ability. World Health Organization (2020). www.who.int/news-room/q-a-detail/ageing-health-ageing-and-functional-ability. Accessed 2 Nov 2020

WHO: Ageing and health (2021). www.who.int/news-room/fact-sheets/detail/ageing-and-health. Accessed 15 Aug 2022

Wiles, J.L., Leibing, A., Guberman, N., Reeve, J., Allen, R.E.S.: The meaning of "aging in place" to older people. Gerontologist **52**(3), 357–366 (2012)

Yalowitz, N., Bassuk, K.: An intergenerational community with supportive services, The NORC model at Penn South Program for seniors. American Society on Aging, San Francisco, CA, USA (1998)

The Effect of Socio-cultural Determinants on Designing Inclusive Telemedicine Tools for the Aging Population

M. Forbes(✉) and J. Chakraborty

Department Computer and Information Sciences, Towson University, Towson, USA
{mforbes,jchakraborty}@towson.edu

Abstract. Telemedicine has served as an effective tool in delivering healthcare to many users via technical modalities. However, non-digital native elderly users who primarily reside in rural communities may naturally feel a digital divide; subsequently having a negative user experience (UX) while navigating these applications. In addition to geographical residence, socio-cultural factors such as age, ethnicity, economic status, and/or education may be variables that cause their limited interaction with technology and lack of receptivity of its adoption. As a result, the functionality of telemedicine applications can pose a challenge without an adequate user interface (UI) design that emphasises their inclusivity. Hence it is important to consider demographic markers that can affect overall usability. The purpose of this paper is to identify and highlight the socio-cultural factors affecting rural elderly users that influence the perceived usefulness of telemedicine applications. Using a grounded theory approach, a qualitative study was conducted using a systematic review of existing literature for data collection to explore relevant usability challenges. Key factors were extracted using a thematic analysis method, with the aim of closing the gap in inclusivity by providing knowledge for further exploratory research – thereby offering opportunities to improve inclusion design for future research and tool development.

1 Introduction

The early phases of COVID-19 were undoubtedly unprecedented times throughout the world. During which period, both physicians and patients experienced a rapid turnaround involving the transition of continuing health services and the reintroduction of telehealth systems. Telemedicine, a subset of telehealth, was utilised within healthcare prior to the global pandemic – it was originally introduced in the United States during the 1960's. Telemedicine is defined as a system of healthcare delivery in which physicians examine distant patients through the use of telecommunications technology (Tan and Ong 2002). While these applications prove beneficial in allowing patients to continue to receive medical services, they also demonstrate some challenges in usability in less urbanised areas. Although telehealth is claimed to increase access to health services for underserved rural populations, there is a general consensus in existing literature that telehealth does not benefit everyone equally, and underserved rural populations are at a particular disadvantage (Svistova et al. 2022). Equally important to their geographical residence, elderly

J. Goodman-Deane et al. (Eds.): CWUAAT 2023, *Design for Sustainable Inclusion*, pp. 50–58, 2023.
https://doi.org/10.1007/978-3-031-28528-8_6

digital immigrants have a perception of risk in using technology due to the absence of a technical background or information technology (IT) training tailored to older users.

Notably, during the transitional period of the stay-at home (lockdown) mandate and deploying telemedicine via information and communication technology (ICT) devices, these individuals were not harmoniously migrated – a particular challenge in rural communities coupled with an increasing aging population and various socio-cultural factors. Per (Tacken et al. 2005), Dutch research reported participants in the age group 64–70, 28% of men versus 8% of women were able to handle a computer. The presence of computers is higher among men, the better educated, higher income and larger households (Tacken et al. 2005). Over 50% of older people experience problems in using a computer and a mobile phone (Tacken et al. 2005). Universal efficacy of any UI is still in its infancy, yet UX researchers can begin creating tools for optimal use across demographics.

The foundation of making a tool optimal for elderly digital immigrants consists of communicating and displaying its projected value. Socio-cultural factors correlated with the aging population are extremely important in the context of telemedicine related to its perceived usefulness (PU) and perceived ease of use (PEU) as we move towards a digital society. It is important to go behind content analysis to investigate users' perceptions of mobile health apps user interfaces (UI) (Alsswey et al. 2018). Given this, UX practitioners should be cognisant of possible age-related health deterioration (e.g., cognition and dexterity) and cultural markers that affect the PEU of future interaction. Poor cultural aspects in UI design regularly mean poor user interaction and hence reduced user acceptance and satisfaction (Alsswey et al. 2018). Therefore, addressing cultural differences of use when designing UIs could improve acceptance, usability and help users to interact in a better way with the interface (Alsswey et al. 2018).

Following decades of research focused on getting older adults to adopt technology, there has not been enough progress to ensure older adults are sufficiently adept to navigate a society in which critical services are increasingly "online only" (Knowles and Hanson 2018). Designers must intermix specific functionality in tool development that will allow elderly digital immigrants in rural communities to engage with the technology as effortlessly as possible. This paper aims to answer the following: R1: Does the adoption of technology affect one ethnicity more than others?

2 Methods

An exploratory research approach was undertaken for data collection by analysing secondary research (n = 53) including case studies, literature reviews, and usability studies employing qualitative methods for sampling, collecting, analysing, and extracting data. Studies were selected by searching scholarly databases. This methodology was selected to further understand the phenomenon of technology adoption by elderly users in rural communities. A qualitative approach was chosen as it emphasises knowledge and understanding of a problem, to provide an adequate synthesis on adapting to technology considering a multitude of factors.

2.1 Search Criteria

The following key terms were used: (Usability OR User Experience) AND (Challenges OR Barriers OR Difficulties) AND (Telemedicine OR Telehealth) AND (Patients OR Clients) AND (Covid-19 Pandemic) AND (Rural Areas OR Rural Communities) AND (Elderly OR Aged OR Older OR Elder OR Geriatric). The results of these searches were screened using an eligibility criteria.

The inclusion criteria for this study included:

- publication written in English;
- full text;
- scholarly peer-reviewed journal articles;
- aging population;
- digital immigrants;
- primary rural residency.

Studies including youth and adolescent age groups, users with sufficient ICT knowledge and settings in urban communities were excluded.

2.2 Study Selection

The selection of studies was completed by using specific search criteria in academic research databases such as ScienceDirect, ACM, Springer, ResearchGate and Journal of Medical Internet Research. Subsequent studies were also selected by reviewing closely related scholarly references citied in primarily selected studies.

2.3 Eligibility Criteria

Qualitative data was extracted from the studies and evaluated based on quality criteria. A checklist was created to further assess the studies and to rule out any discrepancies as illustrated in Table 1.

Table 1. Qualitative research criterion assessment

Quality criterion assessment	Met criteria	Did not meet criteria
Does this study focus on qualitative research?	48	5
Is the qualitative approach clearly identified?	50	3
Is the approach clearly stated?	37	16
Is the sampling method clearly explained?	41	12
Is the method data collection clearly explained?	49	4
Is the method of data analysis free of bias or obstruction?	53	x
Are the findings supported by adequate evidence?	50	3

2.4 Data Extraction and Analysis

Data extraction was conducted independently via a methodological quality assessment based on inclusion criteria, and its validity established by cross examining the content on specific quality criteria for eligible studies. Extracted data includes thematic analysis to further understand how elderly users in rural settings interact with and adapt to technology across ethnicities. A qualitative synthesis was conducted using the Technology Acceptance Model (TAM) framework, providing a grounded theory to better understand the adoption rate of users. This conceptual model helps to define a theoretical basis whether technology will be accepted by users based on two primary factors: perceived usefulness and perceived ease of use. While cross examining socio-cultural factor markers, themes began to emerge – indicating data saturation as illustrated in Fig. 1.

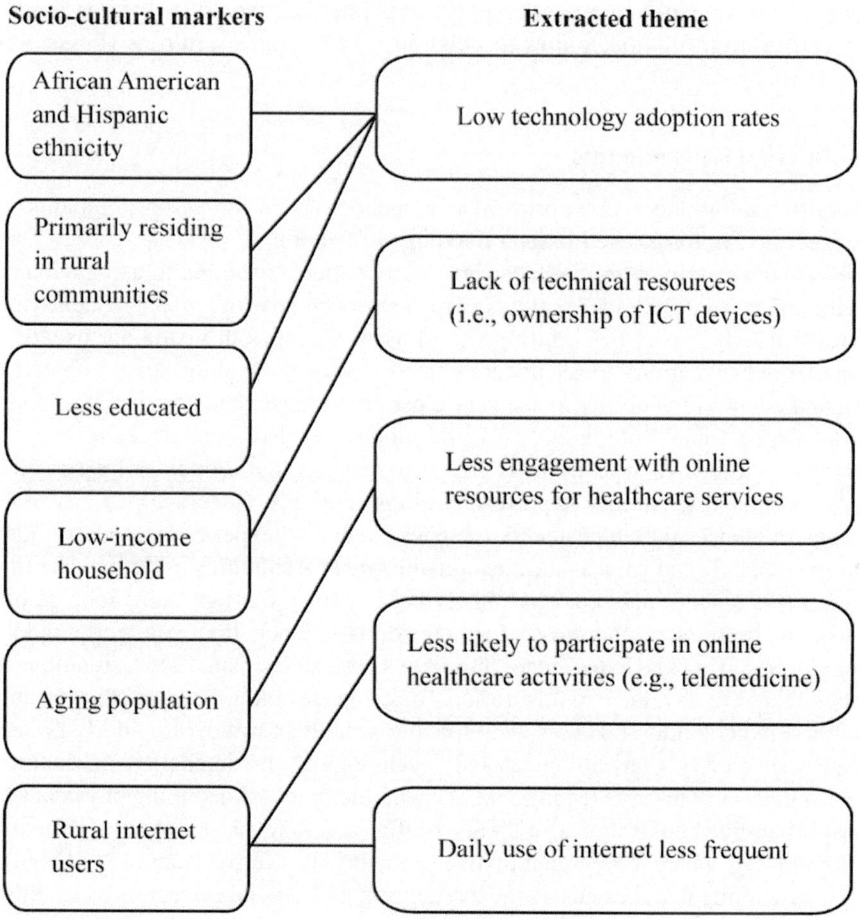

Fig. 1. Thematic analysis chart

3 Thematic Findings

Thematic findings are presented in two categories: Internal and External Determinants, based on socio-cultural factors found in digital immigrants within the aging population in rural communities. Using an exploratory research method, a systematic review of literature established cultural determinants that accompany a projected UX. Internal and external sociocultural factors include health, age, cultural beliefs, geographical residence, social influence and education. Data on the adoption and use of technology such as computers have generally shown that a number of factors, such as education, socioeconomic status, attitudes toward technology, the perceived benefits of technology, and access to technology, influence technology adoption (Czaja et al. 2006). Common reasons given as to why older people report that they do not go online include costs, lack of skills, lack of interest and concerns about security of information (Czaja et al. 2006). Understanding this information early in the design process helps UX practitioners to create a positive experience. An effective user interface would be created based upon perspectives from the disciplines to maximise the overall usefulness (Punchoojit and Hongwarittorrn 2017).

3.1 Internal Determinants

As digital immigrants, perceptions of new technology in the aging population within rural settings are formulated based on a range of factors.

Health tends to be a crucial factor in elderly patients adapting to using telemedicine services. For example, older patients may experience hearing, visual or dexterity difficulties that will impact PEU during use of apps. Changes in vision are usually in the form of a reduced ability to see distant objects, a loss of the ability to see objects to one side, and a loss of the ability to see very close (even faces) and some colors (peripheral vision) (Kececi and Bulduk 2012). Rural elders are also more likely to have certain chronic conditions such as arthritis and hypertension that reduce self-assessed health status and impair the ability to perform various activities (Rogers 2002).

Age influences the older generation because of past experiences, the origination of habits and beliefs and values – influencing the rate of technology adoption. Having lived longer lives, older people are more likely to have lived through more spheres of experience, and more reconfiguration of spheres of experience, than younger people: many family or friend-related experiences, numerous professional situations, situations related to social life, events related to leisure activities, diverse cultural events (Zittoun and Baucal 2021). Age is often taken as a significant variable for studying elderly populations because the elderly's physiological and psychological characteristics are distinct from their younger counterparts (Meng et al. 2020). Elderly users' declining physical and cognitive capabilities possibly cause them to suffer a high level of technology anxiety and health anxiety, which discourage them from using innovative technology (Meng et al. 2020). Assuming that these users must recognise a beneficial impact of ICT – otherwise it is to be expected it will be underutilised voluntarily.

Education and experience might provide additional insights into ICT acceptance and usage among such a highly heterogeneous population group as older adults (Macedo 2017). Traditionally, elderly users learn through human interaction rather than reading books or taking exams (Sergio et al. 2007). Learning styles may differ across ethnicities resulting in disparities in IT adoption. According to Felder and Silverman's active vs. reflective learning dimension, African American learners were clearly reflective, while Caucasian learners were clearly active (Qureshi and York 2008). In terms of sequential versus global understanding, African Americans and Native Americans were clearly sequential, while other ethnic groups showed no significant preferences (Qureshi and York 2008). Given this, preferred learning styles are likely to affect IT adoption across demographics. This experience also shows that the social relationships established between the elderly and their social circles (namely, other adults, children and grandchildren) are important in the motivations and perceptions of elderly people for learning to use ICT (Sergio et al. 2007). With this in mind, it's imperative to incorporate users' learning styles to help them to learn efficiently.

3.2 External Determinants

Social influences, environmental stressors, geographical residency and the accessibility of healthcare and healthcare systems play a critical role in the adoption of IT in cultures across the world. Research has shown a slow rate of IT adoption rates in the United States, with generally less adoption in settings containing high proportion of minorities (Lopez et al. 2011). For example, in the case of Electronic Health Record (EHR) adoption, only 1.5% of hospitals in the United States have a comprehensive EHR system, and, depending on the definition used, between 8% and 12% of hospitals have a basic system (Lopez et al. 2011).

Telemedicine services ideally have a different implementation approach based on geography. PU will be affected by where users reside due to the culture of these environments, as well as readily accessible resources. For instance, urbanised elderly patients may be less likely to resist ICT based on technical resources available and literacy. However, elderly patients in underserved rural settings may need more persuasion because they are not widely or equally exposed to IT on a daily basis or in routine use. Evidence shows decreased access to and use of the Internet for health-related information across age and social and ethnic strata (Lopez et al. 2011). Persons who are older or poor, belong to a racial/ethnic minority, or have a low education level are less likely to use the Internet (Lopez et al. 2011). Pew Hispanic Survey demonstrates that from 2006 to 2008, Internet use among Latino adults increased from 54% to 64% (versus 61% to 63% for blacks and 72% to 76% for whites) (Lopez et al. 2011).

Using telemedicine applications can decrease the travel burden both for providers and families and allows providers to reach almost all rural areas since interactive televideo is already commonly used there for learning purposes (Hosseini and Yilmaz 2019). Rural elderly users, who are digital immigrants, are more likely to decline in physical and mental health because the environmental settings and lack of accessible resources. Environmental settings and social systems affect personal behavior by impacting one's biological condition, and interacting with physiological processes to generate mood states that bias cognition and constrain social interaction (Cole et al. n.d.). In addition, technical resources such as internet and bandwidth are crucial components of ICT integration and utilisation.

Elderly users are most influenced through sociocultural factors – ranging from significant others, family members, church communities and neighbors who can increase the chances of technology adoption. Social networks can consist of multiple circles, with users particularly valuing the emotional and instrumental support of individuals in their closest circle and wanting to share the decision-making with them (Doekhie et al. 2020).

4 Discussion and Conclusion

The aim of this paper was to more deeply explore different factors and barriers to help develop more inclusive healthcare tools for the aging population based in rural settings. With the rapid growth of ICT functionality, current state healthcare tools provide a foundation to optimise inclusion via socio-cultural driven system performances. Adapting to technological advancements can seem a challenge to some users who do not have a sufficient IT background. By further examining social determinants, the survey reached a better comprehension of variables such as health, geographical residence, social influence, demography and income in relation to IT adoption.

A systematic review was conducted to explore IT adoption in older users. Based on thematic analysis, synthesising academic literature and integrating the Technology Acceptance Model (TAM) framework, both internal and external social determinants support in PU and PEU of rural elderly users in ICT devices and telemedicine applications. The framework involving the acceptance of technology was developed to predict individual adoption and use of new information technologies (IT) (Venkatesh and Bala 2008). It posits that individuals' behavioral intention to use an IT is determined by two beliefs: perceived usefulness, defined as the extent to which a person believes that using an IT will enhance his or her job performance and perceived ease of use, defined as the degree to which a person believes that using an IT will be free of effort (Venkatesh and Bala 2008). Older adults in rural areas may deliberately choose to resist using technology as they do not have readily available resources to help them get the skills to use ICT devices, as compared to other demographics.

Using grounded theory analysis to better understand the research question (discussing whether the adoption of technology affects one ethnicity more than others), one can derive that there are disparities in IT adoption across ethnicities. Due to the perception of IT, the adoption rate will vary across users as some embrace the advancements based on their way of life, while others adopt them based on cultural significance. Based on learning styles and access to literary resources, African American and Hispanic ethnicities have a lower adoption rate compared to other ethnicities.

4.1 Limitations

This study reviewed the adoption of technology in rural areas within the aging population using only qualitative studies and literature. A systematic review of peer reviewed literature was conducted rather than using primary methods of data collection.

4.2 Future Work

While disparities may increase because of access inequalities, more research is needed to better understand the barriers to IT use with the aim of creating an adequate design that will improve health ICT use in vulnerable populations. Coupled with technical advancements, existing technology can aid in designing a more inclusive tool based on focusing on social determinant barriers. To illustrate, innovative quantitative techniques could be useful to track unexpected information about psychophysiological (e.g., eye-tracking, heart-rate, galvanic skin response, non-verbal communication) responses of the users to assess their affective and cognitive reactions (Tuena et al. 2020). Eye tracking technology can also be used for evaluating usability and adherence (good > 80%) by using time spent, number of log-ins, or interaction modality, giving additional quantitative data (Tuena et al. 2020). Cultivating technology in ICT can aid in reducing dexterity difficulties while performing commonly overlooked tasks such as ability to turn sound and video on/off, modify screen brightness or adjust screen size.

References

Alsswey, A., Umar, I.N.B., Bervell, B.: Investigating the acceptance of mobile health application user interface cultural-based design to assist Arab elderly users. Int. J. Adv. Comput. Sci. Appl. **9**(8) (2018)

Cole, G., Holtgrave, D., Ríos, N.: Internal and external factors that encourage or discourage health-relevant behaviors. Oak Ridge Institute for Science and Education (n.d.). www.yumpu.com/en/document/read/8553343/internal-and-external-factors-that-encourage-or-discourage. Accessed 25 Jan 2023

Czaja, S.J., Charness, N., Fisk, A.D., Hertzog, C., Nair, S.N., et al.: Factors predicting the use of technology: findings from the Center for Research and Education on Aging and Technology Enhancement (CREATE). Psychol. Aging **21**, 333–352 (2006)

Doekhie, K.D., Buljac-Samardzic, M., Strating, M.M., Paauwe, J.: Elderly patients' decision-making embedded in the social context: a mixed-method analysis of subjective norms and social support. BMC Geriatr. **20**, 53 (2020)

Hosseini, H., Yilmaz, A.: Using telehealth to address pediatric obesity in rural Pennsylvania. Hosp. Top. **97**, 107–118 (2019)

Kececi, A., Bulduk, S.: Health education for the elderly. Geriatrics, pp. 153–177 (2012)

Knowles, B., Hanson, V.: The wisdom of older technology (non)users. Commun. ACM **61**(3), 72–77 (2018)

López, L., Green, A., Tan-McGrory, A., King, R., Betancourt, J.: Bridging the digital divide in health care: the role of health information technology in addressing racial and ethnic disparities. Joint Comm. J. Qual. Patient Saf. **37**, 437–445 (2011)

Macedo, I.M.: Predicting the acceptance and use of information and communication technology by older adults: an empirical exam. Comput. Hum. Behav. **75**, 935–948 (2017)

Meng, F., Guo, X., Zhang, X., Peng, Z., Lai, K.H.: Examining the role of technology anxiety and health anxiety on elderly users' continuance intention for mobile health services use. In: Proceedings of the 53rd Hawaii International Conference on System Sciences, Maui, HI, USA, 6–10 January 2020

Punchoojit, L., Hongwarittorrn, N.: Usability studies on mobile user interface design patterns: a systematic literature review. Adv. Hum. Comput. Interact. **2017**, 1–22 (2017)

Qureshi, S., York, A.: Information technology adoption by small businesses in minority and ethnic communities. In: Proceedings of the 41st Annual Hawaii International Conference on System Sciences, Waikoloa, HI, USA, p. 447, 7–10 January 2008

Rogers, C.: Rural health issues for the older population. Rural Am./Rural Dev. Perspect. **17**, 30–36 (2002)

Sergio, S., Patricia, S., Maite, G., Míriam, A., Laura, L.: Meeting educational needs of the elderly in ICT: two exploratory case studies. Crossroads **14**(2), 1 (2007)

Svistova, J., Harris, C., Fogarty, B., Kulp, C., Lee, A.: Use of telehealth amid the COVID-19 pandemic: experiences of mental health providers serving rural youth and elderly in Pennsylvania. Adm. Policy Mental Health **49**, 530–538 (2022)

Tacken, M., Marcellini, F., Mollenkopf, H., Ruoppila, I., Szeman, Z.: Use and acceptance of new technology by older people. Findings of the international MOBILATE survey: enhancing mobility in later life. Gerontechnology **3**, 126–137 (2005)

Tan, L.T.H., Ong, K.L.: The impact of medical technology on healthcare today. Hong Kong J. Emergency Med. **9**, 231–236 (2002)

Tuena, C., Pedroli, E., Trimarchi, P.D., Gallucci, A., Chiappini, M., et al.: Usability issues of clinical and research applications of virtual reality in older people: a systematic review. Front. Hum. Neurosci. **14**, 93 (2020)

Venkatesh, V., Bala, H.: Technology acceptance model 3 and a research agenda on interventions. Decis. Sci. **39**, 273–315 (2008)

Zittoun, T., Baucal, A.: The relevance of a sociocultural perspective for understanding learning and development in older age. Learn. Cult. Soc. Interact. **28**, 100453 (2021)

Proposing a Systematic Assessment Tool for Evaluating the Architectural Variables of Dementia-Friendly Design in Nursing Homes

T. Golgolnia[1(✉)], M. Kevdzija[2], and G. Marquardt[1]

[1] Chairs of Social and Health Care Buildings and Design, Faculty of Architecture, Technische Universitaet Dresden, Dresden, Germany
{tahere.golgolnia,gesine.marquardt}@tu-dresden.de

[2] Institute of Architecture and Design, Faculty of Architecture and Planning, TU Wien, Vienna, Austria
maja.kevdzija@tuwien.ac.at

Abstract. The world's aging population is growing rapidly; consequently, the population with dementia is also significantly increasing. The relationship between the built environment and Health and Care Outcomes (HCOs) of people with dementia has been established by a considerable amount of research on dementia-friendly design. To implement the findings of studies on dementia-friendly design, an assessment tool of Architectural Variables (AVs) is required to evaluate the effectiveness of the design of built environments, such as healthcare facilities for people with dementia. Despite the existing assessment tools and studies conducted in the field of dementia-friendly design, systematically evaluating the AVs is not possible in practice, because of some challenges discussed in this paper. To address these challenges, this literature-based paper examines the capabilities of an assessment tool (toolkit) to systematically evaluate the dementia-friendly design of nursing homes. For this, three main aspects of the assessment tool including (a) the development base, (b) the users, and (c) the capability of being kept up to date are investigated to find out how they could contribute to the creation of a systematic assessment tool. This paper also proposes a methodology for the development of an assessment tool in three steps. First, the AVs and HCOs, and their interactions should be organised in a database. Second, a structured database of data collection methods and the available data should be developed. Third, the development of analytical relations should be done to provide assessments.

1 Introduction

Dementia is an age-related syndrome that leads to generalised deterioration in cognitive function that causes functional impairment including resulting disability in people's daily lives (Plassman et al. 2007). It is beyond what might be expected from the usual consequences of biological aging (WHO 2021). Dementia affects memory, thinking, orientation, comprehension, calculation, learning capacity, language, and judgment. With the aging of society, numbers with dementia are increasing. It is projected that the number of people with dementia in the world will increase from 55 million at the current

J. Goodman-Deane et al. (Eds.): CWUAAT 2023, *Design for Sustainable Inclusion*, pp. 59–69, 2023.
https://doi.org/10.1007/978-3-031-28528-8_7

time to 139 million in 2050, and the cost annually imposed on societies will reach US$ 2.8 trillion in 2030 (Hoffmann et al. 2020; WHO 2021).

Also, the fact that the consequences of dementia affect other groups, including families, caregivers, and society as a whole, makes it essential to address the challenges linked to people with dementia (Waller and Masterson 2015). On the other hand, evidence on the role of the designed and built environment in supporting people with dementia has been steadily increasing for almost 40 years (Calkins 2018). The physical environment is a critical component of a therapeutic setting for such people (Chaudhury and Cooke 2014; Zeisel et al. 2020, 202) to help them achieve their full potential and avoid unnecessary disability (Davis et al. 2009).

Recently, there has been a much-needed increase in developing detailed and comprehensive design tools for design professionals and healthcare providers. Also, many empirical studies have been carried out that examine the effectiveness of physical environmental features to support the quality of life for people with dementia in nursing homes (Zeisel et al. 2020). Specifically, the subject of empirical studies on dementia by researchers in the field of architecture is the examination of the "Architectural Variables" (AVs) of the built environment on "Health and Care Outcomes" (HCOs) of people with dementia. AVs are those items of the built environment that the architect plays a key role in determining (e.g., scale, density, building layout, and environmental attributes) (Marquardt et al. 2014; Calkins 2018; Chaudhury et al. 2018). HCOs are those outcomes related to various human aspects, including health (physical and mental health), behavior, social interactions, and human identity (Kar and Sarkar 2017; Büter and Marquardt 2020).

By now, it has been established that the built environment in which people with dementia live and are cared for plays a key role in aspects, including their well-being, function, social abilities, and care outcomes (Marquardt et al. 2014). However, we argue that this potential of the built environment could be better used if the outputs of empirical studies were not only the subject of various literature reviews but organised into a systematic assessment tool. It is worth mentioning that although architectural quality cannot be achieved through any assessment tool, the tools can still help architects to make informed design decisions for the benefit of people with dementia.

2 Definition of "Dementia-Friendly Design"

To date, there is no precise definition of the term "dementia-friendly design" (Kirch et al. 2018). However, some efforts to define the underlying concept were made in recent years. The usage of the term "dementia-friendly" may vary, but in general, it refers to the "social" health of people with dementia versus their "medical" health. Defining attributes of dementia-friendliness includes empowerment, aspiration, self-confidence, contribution, participation, meaningful activities, wayfinding ability, sense of safety, accessibility of local facilities, social acceptance, and human rights (Olson and Albensi 2021). The authors (Hebert and Scales 2019) reviewed the literature and identified various applications of the term "dementia-friendly" such as dementia-friendly initiatives, dementia-friendly communities, and dementia-friendly design. To give an example: the concept of "dementia-friendly communities£ refers to cities and communities that cultivate an understanding of dementia and support people with dementia to live well (Downs 2013). As another example, "people with dementia have the same rights as everyone else to be treated with dignity and respect, to lead independent, autonomous lives, and to continue to be active citizens in a society whose opinions are heard and acted upon" (Bartlett 2016).

The concept of dementia-friendly design has been discussed in studies of dementia-friendly initiatives in the environment of long-term care facilities (Hebert and Scales 2019). Also, a set of design principles for dementia-friendly design has been proposed. In most research sources, the principles proposed by Marshall and Fleming–Bennett are mentioned (e.g., providing a human scale, creating a familiar place, and enhancement of visual access) and have been extracted and organised around a set of environmental design principles (Marshall 2011; Zeisel et al. 2020). Generally, although elements and principles of dementia-friendly design have been clearly articulated in the literature, an overarching definition has been conspicuously absent. Such a definition should highlight the importance of the experiences of people with dementia and the need to consider the organisational, social, and physical environments that influence those experiences. Hence, a dementia-friendly environment could be defined as a cohesive system of support that recognises the experiences of people with dementia and provides assistance for them to remain engaged in everyday life (Barbosa et al. 2017).

3 Role of the Assessment Tool for Evaluating the AVs of Dementia-Friendly Design

Although there is a considerable number of empirical studies which are the primary resource for dementia-friendly design principles, to our knowledge they have not been organised to be used effectively. Also, the lack of a cohesive definition of dementia-friendly design could lead to various interpretations of its principles. It should also be considered that the research base of dementia-friendly design in long-term care facilities is more robust than that of acute care, so dementia-friendly specific studies require more consistent use of evidence-based assessment tools, with reporting of psychometric properties (Hebert and Scales 2019). Given the nature of an assessment tool, as well as the features it should have, an assessment tool for evaluating the AVs in dementia-friendly design could play a role as both a definer and an organiser of the principles involved.

Assessment means the process of testing and making a judgment about (Cambridge Dictionary 2022) the aspects of an environment to improve it for the purpose for which it was created. Assessment involves three steps: identifying influential variables, measuring them, and collecting and analysing data on the measures needed to achieve a purpose (e.g., to specify design defects or to test theory) (Carp 1994). An assessment tool is based on the research evidence used for evaluating the built environment's design. The assessment tool focuses on those aspects of the physical environment known to impact people with dementia. It assesses not only the physical environment (such as floor coverings and the use of paint colors) but also the way that the environment encourages people to behave and interact (Waller and Masterson 2015). In all three steps mentioned above assessment tools use various types of information, including residents' views, ratings by assessors, and technical measurements. Whether measurements are single items or sophisticated scales, either subjective or objective, they should assess the environment distinctly, not confounding it with other influences on outcomes under study (Carp 1994). Comprehensive assessment is essential in providing appropriate care facilities and improving the quality of life for people with dementia. This assessment tool will be suitable for use by researchers in the field of architecture in the process of evidence-based design. It will also be useful for architects in the planning process.

According to the discussion above about an assessment tool, there are two questions:

- Q1: How many and what kind of features and requirements for evaluating the AVs of dementia-friendly design have been included in the existing assessment tools?
- Q2: How should the shortcomings of the existing assessment tools be addressed, and what shortcomings should be prioritised?

A systematic review of existing assessment tools is carried out to answer these questions. To find out how specifically they evaluate the impact of the environment, particularly the impact of AVs on people with dementia, in long-term care facilities. In the following section, the method for conducting a systematic review is presented first, followed by the results, and finally by a methodology for developing an efficient assessment tool.

4 Method

In identifying and analysing existing assessment tools, inclusively and exclusively dementia-focused, two fundamental considerations have been considered.

First, studies have been selected to give maximum coverage of pre-existing ones. For this purpose, two complementary systematic literature reviews covering the period from 1990 to 2021 were used to identify assessment tools. The first review by (Elf et al. 2017) included publications from 1990 to 2016 and focused on a systematic review of environmental assessment tools for assessing the quality of the physical environment in health care. The second systematic review study conducted by Calkins et al. (2022) reviewed the period from 2017 to 2021 and focused on evaluating environmental assessment tools for long-term care facilities for people with dementia. The search terms (tool, instrument, scale, assessment, measurement, evaluate, screening, the physical healthcare environment, healthcare space, healthcare setting, hospital, healthcare architecture, healthcare building, healthcare design) and the search databases (Medline, CINAHL, PsycInfo, and Avery Index) were identical in both studies (Elf et al. 2017; Calkins et al. 2022).

Second, a review of the identified assessment tools based on the following criterion has been done. The criterion determines the degree of relevance of each assessment tool to the aim of evaluating the AVs. Then, the existing assessment tools are sorted so that the ones most relevant to AVs are prioritised for analysis to answer the questions raised in Sect. 3. The next section presents the chosen assessment tools and their analysis.

5 Result of Environmental Assessment Tools Analysis

From the previously mentioned assessment criteria, 13 assessment tools were identified, which are presented in Table 1. These tools are sorted based on the degree of relevance of items in an assessment tool to the AVs. This criterion is described as R(i) in Eq. 1. In this relationship, TIs(i) is equal to the total number of items that are considered in a particular assessment tool (here i), and RIs(i) is the number of items that are related to AVs.

$$R(i) = \frac{RIs(i)}{TIs(i)} * 100\% \tag{1}$$

R(i) as a criterion for sortation is crucial since the existing assessment tools consider items, some of which are unrelated to AVs. Thus, the greater the percentage is, the more an assessment tool relates to AVs, thus making it more relevant for architectural research.

Table 1. The number of items allocated to AVs in environmental assessment tools

R(i)		Tool's acronym	Tool's name
Level 1 0% to 50%	0%	PEAP	Professional Environmental Assessment Protocol
	0%	EOH	Experience of Home
	10%	NURS	Nursing Unit Rating Scale
	18%	ACC-2.0 NH	Artifacts of Culture Change 2.0
Level 2 51% to 80%	55%	SCEAM	Sheffield Care Environment Assessment Matrix
	60%	EQuAL	Environmental Quality Assessment for Living
	61%	DEAP	Dining Environment Audit Protocol
	68%	EVOLVE	Evaluation of Older People's Living Environments
	70%	EA-DDAT	Environments for Ageing and Dementia Design Assessment Tool
	78%	E-B Model	Environment-Behaviour Model
Level 3 81% to 100%	86%	EAT	Environmental Assessment Tool
	86%	EHE-EAT	Enhancing Healing Environment - Environmental Assessment Tool
	90%	TESS-NH	Therapeutic Environment Screening Survey for Nursing Homes

Considering the R(i) amounts, assessment tools are classified in three ranges including 0%–50%, 51%–80% and 81%–100%. The assessment tools in the 81%–100% percentage range incorporate more AVs than any of the other tools. That is why they have been used as the most representative examples of environmental assessment tools (Table 2).

On investigating the existing assessment tools, the answer to question 1 in Sect. 3 can be resolved into three main problems, whose resolution could result in more effective assessment tools.

Table 2. Environmental assessment tools suitable as an analytical basis for evaluating the AVs of dementia-friendly design

Tool's acronym	Individual tool features
EHE-EAT	• Developed for assessing hospitals and health centers • Direct observational tool • Divided into seven criteria that the environment should promote: meaningful interaction between patients, their families, and staff; well-being; eating and drinking; mobility; continence, and personal hygiene; orientation; and calm, safety, and security (Waller and Masterson 2015)
EAT	• Developed for assessing hospitals and residential aged care facilities • Direct observational tool • Consists of 72 questions grouped into ten subscales: safety; size; visual access; reduction of unnecessary stimuli; highlighting of useful stimuli; provision for wandering and outdoor areas; familiarity; privacy and community; community links; and domestic activities (Smith et al. 2012)
TESS-NH	• Developed for assessing nursing homes • Direct observational tool • Divided into 84 discrete items plus one global item that covers 13 domains, including exit control; maintenance; cleanliness; safety; orientation/cueing; privacy; unit autonomy; outdoor access; lighting; noise; visual/tactile stimulation; space/seating; and familiarity/homeliness (Sloane et al. 2002)

a. The basis of assessment tool development: The lack of the comprehensiveness needed to cover a wide range of AVs that affect the HCOs in dementia-friendly design could be considered the development base's main problem. This problem results from the lack of a systematic classification of AVs for assessment, which explains why only some of the AVs in a not well-developed classification combined with some criteria and principles, are available in the existing assessment tools. These problems have caused other negative side effects. The existing assessment tools do not consider the cumulative interactions and correlated impact of AVs on HCOs. As a result, the existing assessment tools provide little to no flexibility for adaptation to a specific architectural research question.

b. The usage and users of the assessment tool: The existing assessment tools are not widely used in architectural practice, which is also due to their lack of a classification of the AVs. Another limitation is the difficulty of using the existing assessment tools in the pre-construction stage (design stage). It is unclear for what type of users and level of expertise most assessment tools were developed. Specifically, to develop an assessment tool to evaluate the AVs of dementia-friendly design in nursing homes, the target users are architects and design researchers.

c. The ability to be updated as a basis for further research: The existing assessment tools are not as extendable and flexible as would be necessary for researchers and architects to better access the required evidence-based inputs for dementia-friendly design. More specifically, health-related variables in dementia-friendly design used in an assessment tool are not accessible through medium such as an online platform, database, or software. Thus, it is a problem for the existing assessment tools to incorporate new research effectively.

Another problem is the difficulty of comparing the results of the tools because they use different variables to assess the built environment. A methodology is presented in the next section to address the problems discussed above, which also answers the second question in Sect. 3.

6 Proposing a Methodology for the Development of a Systematic Assessment Tool for Evaluating the AVs of Dementia-Friendly Design

A methodology is proposed to develop a systematic assessment tool that addresses the shortcomings of the existing ones. This methodology consists of three steps, as shown in Fig. 1. The steps are (1) comprehensive classification of AVs and HCOs as the first step, (2) creation of organised databases of data collection methods as the second step, and (3) defining and extracting analytical relations whilst emphasising quantitative relations.

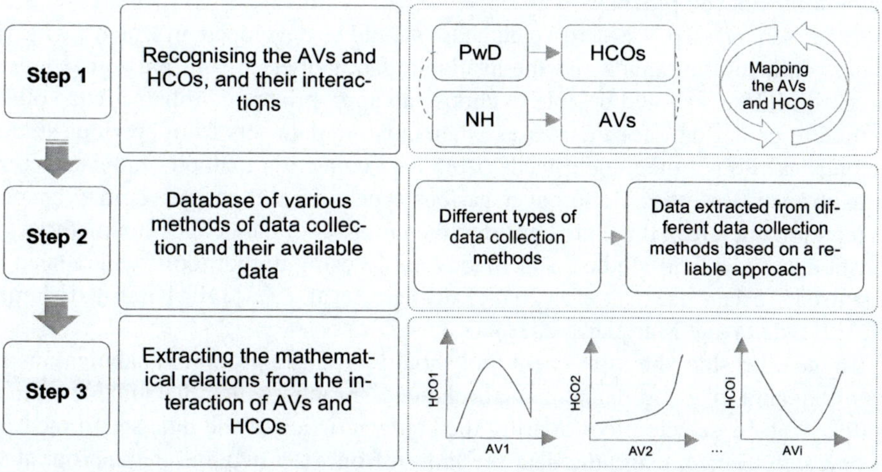

Fig. 1. The steps for the development of a systematic assessment tool for evaluating the AVs of dementia-friendly design, (PwD = People with Dementia, NH = Nursing Homes, AVs = Architectural Variables, HCOs = Health and Care Outcomes)

In the first step, which is the initial basis of the assessment tool development, the AVs, and HCOs, as well as their interactions and correlation with each other, will be recognised and organised in a database. The more AVs and HCOs are categorised and analysed in this step, the more comprehensive and reliable the assessment tool will be. AVs and HCOs will be recognised and gathered mainly through the existing literature reviews and the previous studies that provide a scientific development base for this assessment tool. Then, interactions between the recognised AVs and HCOs need to be identified, especially focusing on interplays originating from unequal and inconsistent effects that an AV could have on different HCOs. In other words, previous studies have considered the relationship between an AV and HCO mainly as a bilateral relation and have used the matrix method for tool development. Through the matrix method, an AV change could change another AV's effectiveness on HCOs. For this purpose, when the effects of an AV on HCOs are investigated, its effect on other AVs, and consequently on HCOs indirectly, should be taken into account.

Due to the behavior of the interactions between the AVs and HCOs, the "system dynamics" method can be considered an alternative to methods previously used. The system dynamics method's effectiveness in simulating complex systems such as healthcare systems has been proved in cases including subtle mutual effects, internal and external effects, and nonlinear behaviors of variable interaction. (Fone et al. 2003; Homer and Hirsch 2006; Forrester 2007) The system dynamics method was developed as an approach to address the challenges of analysing systems that result from the complexities in their behavior (Mustafee et al. 2010). The main advantages of the system dynamics method include its ability to identify system variables accurately, recognise linear and nonlinear interlinkages, examine the long-term effects of decisions on the system, and raise awareness of the nature of problems (Allahi et al. 2018). These advantages make the system dynamics method a potential solution for the first step of the proposed methodology development.

In the second step, a structured database should be developed, in which a wide range of data collection methods and the available data gathered from previous studies are categorised. Users should be able to choose an appropriate and reliable data collection method to gather data for their cases when existing data sets from previous studies in the database do not meet their needs. Both data collection methods and data sets must be as extensive as possible to cover various types of cases that users may encounter. To mention but a few, there are considerable numbers of data collection methods, from questionnaires, cohort studies, and shadowing to using digital tools, which need to be organised in a database (Luz et al. 2018; Darlington et al. 2021; Julia Kirch and Marquardt 2021; Kevdzija and Marquardt 2021).

In the third step, the assessment tool should provide a group of analytical methods, mainly quantitative and mathematical relations, through which the suitable method of analysis can be selected, considering the characteristics of the data set formed in the second step. Hence, using the data set chosen from step two and the appropriate data analysis method from step three, the effects of AVs in a specific design case could be analysed quantifiably. Then decisions will be made on a solid base and systematic approach. This approach, also in line with previous steps, could address the lack of consensus among users around an assessment tool, and as a result boost its use in practice. Also, when an assessment tool is used, it would be more able to be kept up to date and consequently enhance its validity, since the most recent achievements in the field of dementia-friendly design would be reflected and included in that.

The steps explained above lead to the formation of the assessment tool. After that, orienting the tool toward a platform should be considered. This would make it easier to connect the databases to users and enable them to participate effectively in updating it through the capabilities included in the platform. To accomplish this, an appropriate infrastructure should be identified as the medium for defining the assessment tool – for example, software. Then the platform's access points are configured to clarify the accessibility of the information of the platform to users as well as the extent to which they could make changes in the platform.

7 Conclusion

Following analysis and recognition of improvement potential in the context of dementia-friendly design, developing a systematic assessment tool is considered a reliable solution to the associated challenges. To this end, the steps required to develop the targeted assessment tool were structured, which will serve as the foundation for ongoing research based on the findings of this study. First, a comprehensive and systematic classification of AVs and HCOs and the analysis of interactions between them needs to be done to cover as many of them as possible. This will likely lead to an acceptable level of consensus between user groups to avoid misinterpretation. Then, as the second step, data necessary to analyse the interactions of AVs and HCOs should be gathered, which requires organised databases to increase their usability. Finally, analytical relations focusing on quantitative types should be developed to provide reliable and meaningful assessments that could result in effective design decision-making with as much consensus as possible.

In terms of anticipated impacts, this assessment tool's main effect is expected to be future research on dementia-friendly design conducted more systematically and based on standard processes. Also, other effects of this assessment tool could concern two groups, policymakers and architects. Policymakers will be enabled to optimally set specific standards and rules, or issue permits to design, build or modernise nursing homes for people with dementia. Architects could rely on this assessment tool to implement dementia-friendly design criteria.

References

Allahi, F., De Leeuw, S., Sabet, E., Kian, R., Damiani, L., et al.: A review of system dynamics models applied in social and humanitarian researches. In: Proceedings of the World Congress on Engineering 2018 (WCE 2018), Imperial College London, London, 4–6 July 2018, pp. 789–794 (2018)

Barbosa, A., Lord, K., Blighe, A., Mountain, G.: Dementia care mapping in long-term care settings: a systematic review of the evidence. Int. Psychogeriatr. **29**(10), 1609–1618 (2017). https://doi.org/10.1017/S1041610217001028

Bartlett, R.: Scanning the conceptual horizons of citizenship. Dementia **15**(3), 453–461 (2016). https://doi.org/10.1177/1471301216644114

Büter, K., Marquardt, G.: Dementia-Friendly Hospital Buildings: Construction and Design Manual. DOM Publishers, Berlin (2020)

Calkins, M.P.: From research to application: supportive and therapeutic environments for people living with dementia. Gerontologist **58**(suppl. 1), 114–128 (2018). https://doi.org/10.1093/geront/gnx146

Calkins, M.P., Kaup, M.L., Abushousheh, A.M.: Evaluation of environmental assessment tools for settings for individuals living with dementia. Alzheimer's Dement. Transl. Res. Clin. Interv. **8**(1), e12353 (2022). https://doi.org/10.1002/trc2.12353

Cambridge University Press: Assessment. In: Cambridge Dictionary (2022). https://dictionary.cambridge.org/dictionary/english/assessment. Accessed 16 Jan 2023

Carp, F.M.: Assessing the environment. Annu. Rev. Gerontol. Geriatr. **14**(1), 302–323 (1994). https://doi.org/10.1891/0198-8794.14.1.302

Chaudhury, H., Cooke, H.: Design matters in dementia care: the role of the physical environment in dementia care settings. In: Downs, M., Bowers, B. (eds.) Excellence in Dementia Care, 2nd edn, pp. 144–158. Open University Press (2014)

Chaudhury, H., Cooke, H.A., Cowie, H., Razaghi, L.: The influence of the physical environment on residents with dementia in long-term care settings: a review of the empirical literature. Gerontologist **58**(5), e325–e337 (2018). https://doi.org/10.1093/geront/gnw259

Darlington, N., Arthur, A., Woodward, M., Buckner, S., Killett, A., et al.: A survey of the experience of living with dementia in a dementia-friendly community. Dementia **20**(5), 1711–1722 (2021). https://doi.org/10.1177/1471301220965552

Davis, S., Byers, S., Nay, R., Koch, S.: Guiding design of dementia friendly environments in residential care settings: considering the living experiences. Dementia **8**(2), 185–203 (2009). https://doi.org/10.1177/1471301209103250

Downs, M.: Putting people - and compassion - first: the United Kingdom's approach to person-centered care for individuals with dementia. Generations **37**(3), 53–59 (2013)

Elf, M., Nordin, S., Wijk, H., Mckee, K.J.: A systematic review of the psychometric properties of instruments for assessing the quality of the physical environment in healthcare. J. Adv. Nurs. **73**(12), 2796–2816 (2017). https://doi.org/10.1111/jan.13281

Fone, D., Hollinghurst, S., Temple, M., Round, A., Lester, N., et al.: Systematic review of the use and value of computer simulation modelling in population health and health care delivery. J. Public Health Med. **25**(4), 325–335 (2003). https://doi.org/10.1093/pubmed/fdg075

Forrester, J.W.: System dynamics—the next fifty years. Syst. Dyn. Rev. **23**(2–3), 359–370 (2007). https://doi.org/10.1002/sdr.381

Hebert, C.A., Scales, K.: Dementia friendly initiatives: a state of the science review. Dementia **18**(5), 1858–1895 (2019). https://doi.org/10.1177/1471301217731433

Hoffmann, J., Sotting, U., Isfort, M., Tucman, D., Weil, K., et al.: The care of acute patients with dementia at the special care unit, "Station Silvia": results of an evaluation study. Rep. Glob. Heath Res. **3**, 125 (2020). https://doi.org/10.29011/2690-9480.100125

Homer, J.B., Hirsch, G.B.: System dynamics modeling for public health: background and opportunities. Am. J. Public Health **96**(3), 452–458 (2006). https://doi.org/10.2105/AJPH.2005.062059

Kar, Z., Sarkar, A.: Exploring the role of environment-behavior studies (EBS) in residential architecture - from literature review to field study. IRJET **4**(11), 409–418 (2017)

Kevdzija, M., Marquardt, G.: Stroke patients' nonscheduled activity during inpatient rehabilitation and its relationship with the architectural layout: a multicenter shadowing study. Top. Stroke Rehabil. **29**(1), 9–15 (2021). https://doi.org/10.1080/10749357.2020.1871281

Kirch, J., Marquardt, G.: Towards human-centred general hospitals: the potential of dementia-friendly design. Archit. Sci. Rev. (2021). https://doi.org/10.1080/00038628.2021.1933889

Kirch, J., Marquardt, G., Bueter, K.: Breaking down barriers: promoting a new look at dementia-friendly design. In: Langdon, P., Lazar, J., Heylighen, A., Dong, H. (eds.) CWUAAT 2018, pp. 125–133. Springer, Cham (2018). https://doi.org/10.1007/978-3-319-75028-6_11

Luz, S., de la Fuente, S., Albert, P.: A method for analysis of patient speech in dialogue for dementia detection. arXiv:1811.09919 (2018). http://arxiv.org/abs/1811.09919. Accessed 16 Jan 2023

Marquardt, G., Bueter, K., Motzek, T.: Impact of the design of the built environment on people with dementia: an evidence-based review. HERD: Health Environ. Res. Des. J. **8**(1), 127–157 (2014). https://doi.org/10.1177/193758671400800111

Marshall, M.: How it helps to see dementia as a disability. J. Dement. Care **6**, 15–17 (2011)

Mustafee, N., Katsaliaki, K., Taylor, S.J.E.: Profiling literature in healthcare simulation. SIMULATION **86**(8–9), 543–558 (2010). https://doi.org/10.1177/0037549709359090

Olson, N.L., Albensi, B.C.: Dementia-friendly "design": impact on COVID-19 death rates in long-term care facilities around the world. J. Alzheimer's Dis. **81**(2), 427–450 (2021). https://doi.org/10.3233/JAD-210017

Plassman, B.L., Langa, K.M., Fisher, G.G., Heeringa, S.G., Weir, D.R., et al.: Prevalence of dementia in the United States: the aging, demographics, and memory study. Neuroepidemiology **29**(1–2), 125–132 (2007). https://doi.org/10.1159/000109998

Sloane, P.D., Mitchell, C.M., Weisman, G., Zimmerman, S., Long-Foley, K.M., et al.: The therapeutic environment screening survey for nursing homes (TESS-NH): an observational instrument for assessing the physical environment of institutional settings for persons with dementia. J. Gerontol. Ser. B Psychol. Sci. Soc. Sci. **57**(2), 69–78 (2002). https://doi.org/10.1093/geronb/57.2.s69

Smith, R., Fleming, R., Chenoweth, L., Jeon, Y.-H., Stein-Parbury, J., et al.: Validation of the environmental audit tool in both purpose-built and non-purpose-built dementia care settings. Australas. J. Ageing **31**(3), 159–163 (2012). https://doi.org/10.1111/j.1741-6612.2011.00559.x

Waller, S., Masterson, A.: Designing dementia-friendly hospital environments. Future Hosp. J. **2**(1), 63–68 (2015). https://doi.org/10.7861/futurehosp.2-1-63

WHO: Dementia. World Health Organization (2021). www.who.int/news-room/fact-sheets/detail/dementia. Accessed 16 Jan 2023

Zeisel, J., Bennett, K., Fleming, R.: World Alzheimer report 2020: design, dignity, dementia: dementia-related design and the built environment (2020). www.alzint.org/resource/world-alzheimer-report-2020/. Accessed 16 Jan 2023

Inclusive Built Environments

Reframing Independent Living: Co-producing Homes to Age-in-Place

C. Resendiz-Villasenor[1](✉), F. Pour-Rahimian[1], and P. Carnemolla[2]

[1] Centre of Construction Innovation and Research, School of Computing, Engineering and Digital Technologies, Teesside University, Middlesbrough, UK
{C.ResendizVillasenor,F.Rahimian}@tees.ac.uk

[2] Faculty of Design Architecture and Building, School of Built Environment, University of Technology Sydney, Ultimo, Australia
Phillippa.Carnemolla@uts.edu.au

Abstract. Independent Living has influenced strategies for ageing in place, impacting public policies regarding housing and care services. However, this approach remains challenging as products and services for older people are stigmatising, reducing adoption levels. This study aims to understand how people would integrate care for their future selves in a home environment via two coproduction sessions (online and in-person). A multi-generational group of participants with lived experience as informal caregivers provided insights about built environment aspects, home and neighbourhood, meaningful relationships and advanced activities of daily living for later life. Data was analysed qualitatively. The outcomes show how coproduction can be used to develop home and care concepts for aging in place, including insights about preferred home configurations and how physical and digital interactions with products available in the market can improve quality of life as people age. This research can be used as a reference for designing inclusive new-builds, home adaptations and assisted-living technologies and interfaces. By using this approach, independent living strategies could evolve to support positive interdependence networks of people to use digital and physical spaces involving and enhancing self-care and informal and formal care services.

1 Introduction

1.1 Rethinking Independent Living

The population age shift is a global concern. In 2021, the World Health Organization (WHO 2020b) declared the start of the Decade of Healthy Ageing, a collaboration to improve older people's lives, families, and communities. Most of the areas of this initiative are directly related to functional ability (Age-friendly environments, integrated care, and long-term care), which refers to the relationship between intrinsic capacity (mental and physical) and environmental factors (home, community and society) (Moreno-Agostino et al. 2021); such environmental factors are perceived by users as barriers or facilitators and have high relevance for people to maintain functional ability, determining the level of age-friendliness of the Built Environment (BE). Functional ability is associated with Independent Living (IL), which refers to the degree to which

J. Goodman-Deane et al. (Eds.): CWUAAT 2023, *Design for Sustainable Inclusion*, pp. 73–82, 2023.
https://doi.org/10.1007/978-3-031-28528-8_8

adults can care for themselves and their property; it can also be understood as a sense of autonomy (Enkvis et al. 2012; WHO 2020a; Lu et al. 2021). Ball and Nanda (2013) mention the importance of IL and its correlation to health and general wellbeing. Peek et al. (2014) found that increased independence reduces the burden on family caregivers, specifically when implementing assistive technologies.

The IL assessment evaluates memory, orientation, money management, home and transportation, health and safety and social adjustment (Anderten-Loeb 2003). It focuses on basic, instrumental and advanced Activities of Daily Living. This last refers to productive activity and leisure-time tasks (WHO 2022) and was introduced less than a decade ago by De Vriendt et al. (2013) to evaluate patients with mild cognitive disorders. Researchers highlight the complexity of assessing the validity and reliability of a scale on advanced activities because the approach is highly individual (Dias et al. 2019).

1.2 Designing for Independent Living

Human-Centred Design has become a more common practice in technology development for assisted living and age-friendly environments. Crews (2022) notes that designing built environments to accommodate and improve human-environment interactions for frail seniors will enhance usability and accessibility for most user groups. On the other hand, assistive technologies have been researched thoroughly in reactive health scenarios (Haux et al. 2014; Hu et al. 2020). Studies focused on non-specialised technologies for assisted living have concluded that further investigation is required; i.e., Lopez et al. (2019) concluded after a heuristic evaluation of smart home technologies that off-the-shelf devices needed further development to support older adults living independently.

Coproduction can be used in product development to ensure alignment with users' needs, including BE and integrated care (Caron 2019). Literature suggests some intersections between coproduction, co-creation, co-design and participatory design. According to Alzheimer's Society (2022), co-creation is about designing with people instead of for them, with the ultimate goal of producing an outcome of higher quality, ensuring that the result is appropriate for individuals and impacts them positively. Wilton (2021) emphasises the change of the discursive approach from 'What is the matter with you?' to 'What matters to you?', creating a bond of empathy when 'seeing those who use services as equals and experts in their own experience'. Pirinen (2016) compared this approach with traditional developments for case studies of senior villages in Finland and highlighted the benefits of including residents in the design logic. Brookfield et al. (2020) used this method to design age-friendly homes and neighbourhoods in the UK. Rogers et al. (2021) used it to study the application of human-robot interactions. The benefits were twofold: developers practised an improved design process, and users showed an increased level of acceptance for the technology.

There are opportunities to incorporate people's lived experiences regarding care into IL solutions that include IoT and BE; however, this is yet to be studied from the point of view of end users who decide to plan for later life. This research answers the following question: How do informal caregivers use their lived experience to adopt age-in-place strategies?

2 Methods

A housing concept (formed by a modular house design and a series of off-the-shelf technologies, objects and systems) is proposed, guided by the Double Diamond Method (Design Council 2007). The method consists of four stages for product/service creation (from ideation to delivery): Discover, define, develop and deliver. The scope of this paper is to describe the 'Develop' phase of the housing concept. Peace (2022) suggests three scales for approaching this type of research; a micro-scale referring to locality; and meso and macro scales related to national and global aspects. For this research, the focus of environmental factors has been defined by a micro-scale, which includes dwelling, community and neighbourhood.

The workshops were delivered online and in-person between Jun and Sep 2022, advertised as 'Co-Production Workshop: Designing for my future self'. Recruitment took place via the Intergenerational National Network Scotland and MyCareMatters Network to reach a wide and diverse audience. The sample consisted of a multi-generational group of participants with lived experience as informal caregivers who provided insights about built environment aspects, home and neighbourhood, meaningful relationships, and activities to be sustained long term. All participants signed an Informed consent and filled out a participant information form (demographics and age-in-place strategy). Workshops lasted ~120 min, audio and video were recorded, and the conversations transcribed. Data were later processed using thematic analysis (Braun and Clarke 2022), and participants' answers were anonymised. This research gained ethical approval from the SCEDT Research Ethics Sub-Committee at Teesside University.

The workshops comprised two main sections: **1. Ice-Breaker:** Participants discussed micro-scale preferences and priorities for their future home, selected a point in a timeline where they positioned themselves in the future, and described the location, house type and neighbourhood they preferred. **2. Coproduction session:** A series of pre-defined 2D modules (Building Blocks) was introduced to the participants, so they could use them to create a house layout for their future selves. They chose between 2 approaches: 1) Drag and drop the Building Blocks into a blank canvas, or 2) Use a home template (4 templates with different arrangements were developed in advance). After the group agreed on a layout, they were introduced to a catalogue with a series of off-the-shelf products (IoT and design elements related to indoor air quality, daylight, thermal comfort, accessibility, and adaptability). The series of products were presented as a wish list to implement in the house designed previously. The wish list had the following prompts: 'must have', 'nice to have' or 'not important' to inform the discussion. Then, the positive interdependence framework (Fig. 1) was discussed, and each participant brainstormed about their preferences for their future, including advanced activities of daily living (meaningful activities), Home, Community and Support Networks, as well as the preferred interactions (Digital or Physical). In this section, participants were asked how their experience with care provision impacted their strategies for ageing in place in later life.

3 Results

This research aimed to understand how people plan for the future and integrate their experiences into a housing solution. Twelve people participated (aged 32–82; 5 women,

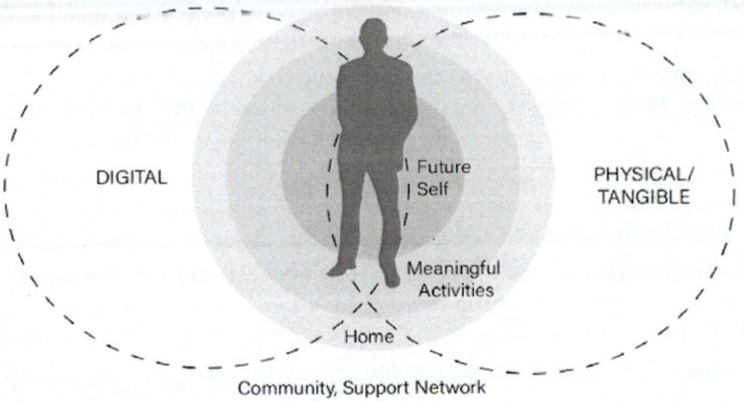

Fig. 1. Positive interdependence framework. Based on Carnemolla (2018)

7 men) in online (7 participants) and in-person (5 participants) workshops. All of them were living in rural locations in Scotland, had a strategy to age in place by home adaptation or newbuild (custom-build) and had lived experience as informal care providers. The results of this work are presented in three sections, describing the micro-scale preferences, technologies and interactions, and how the care experience is transferred into participants' strategies.

Fig. 2. Layout intervention outcomes from online session using a digital collaboration platform (Miro.com) (Left); In-Person Session (Middle); Interventions with Post-it notes over the Positive Interdependence Framework prints (In-person Session) (Right).

3.1 Micro-scale Preferences for Place Attachment

There was a general awareness of the challenges of living in a rural area and how this could affect participants' lifestyles in the long term. Themes have been divided into micro-scale components: community, neighbourhood, and dwelling.

Community and Neighbourhood: Participants living in accessible small towns mentioned that, if required, they could use the bus to get groceries and continue activities in their communities: "*I kinda like the isolation, but I also like to have that choice to be able to go home, come back, and there's still having another community contact... human contact [...] sometimes living in a block of flats in the centre of a big city can be just as lonely, cause you don't know your neighbours, you don't speak to your neighbours and you don't see them.*" P6, 57, Female.

Dwelling: Place attachment was discussed, as well as home adaptations. Participants living in buildings with stairs showed more concern about their strategies. The installation of a stairlift was prompted in the conversation; however, participants did not show interest: "*I live in an upstairs flat, so that also has stairs...It would be... it's good exercise for me, but it might become challenging.*" P1, 82, Female. Others explained how they moved a bungalow to facilitate moving around. When discussing the type of dwelling they preferred for later life, all the participants mentioned they would prefer to live in a detached building to avoid noise from neighbours and increase the sense of privacy. The Building Blocks showed how to produce a home layout, and 4 layouts were provided. Participants in both workshops preferred using pre-defined layouts. In both workshops, 2 double bedrooms were added, indicating the need to allocate space for visitors (Fig. 2 Shows the addition of 2 bedrooms into one of the layouts, as well as the relocation of the mobility storage). Another area that kept participants engaged in the discussion was the bathroom/wet room. One group decided to keep the bath, and the other one to change it to a wet room layout.

3.2 Technologies and Interactions

The dataset describing technologies and interactions was coded, and the results have been divided into technology reliance and reluctance themes to map the polarised ideas discussed during the workshops.

Technology Reliance: Some participants discussed its potential uses for their future self, presenting possible scenarios and the interactions expected in case of an event: "*Having a watch* [points at smartwatch on wrist], *it's something I'm already used to wearing. So I think I could continue using it and in case of emergency, fall down the stairs or something like that, I could reach to an emergency or to contact someone for help.*" P3, 32, Male. Having experience with telecare for physical and mental health prompted others' perspectives regarding the use of digital medical records: "*I have some health records on my phone, some of my medical history... If something was to happen to me, they could get into my phone, and they would see that stuff.*" P2, 47, Female; and virtual reality: "*I have a 'remote care' network already, and that's why I make time for online activities because if someone is not there, the question comes thrown up. In the groups that I take part in, quite a number of people are disabled, so we always have that in mind and say: 'Is anyone okay?', we just want to make sure... If someone doesn't reply or just appears to not being there, there's like 'Aha. Somebody make sure just to double check with that person'.*" P8, 33, Male.

Technology Reluctance: Some concerns included security and privacy: "[...]*think about how secure are these systems? What can I say... I'm thinking of 'Big Brother-ish'*" P1, 82, Female; and concerns about technology advances affecting cognitive health: "*My fear is that it will make people less visible, less active and that our brains have to think less.*" P10, 63, Male. Other technology reluctance aspects were related to price or perceived value: "*It's one of those things that if you have the extra budget for it, it's worth it.*" P9, 32, Female. Another participant shared his preference for digital applications to enhance smartphone capabilities instead of needing additional sensors, i.e. wearables: "*I would rather have the apps in my phone because they tend to be free and cheap-ish*" P8, 33, Male.

3.3 Transferring Informal Care Experiences

Some of the care experiences participants had had in their lifetimes showed an impact on their actions and on how they would potentially use technology: "*I keep tracking my steps, and unless I'm short on steps before I take my shower in the evening, I can run in one spot to add steps. You don't need to go out, and you don't need lots of space to do physical activity. Even do stretches, as my mum did since she was much younger, and she is still active and independent. I think that's the mentality I want to have*" P10, 63, Male. Another participant describes her experience of how care provision has been supported by family members and the concerns about family medical records: 'My mother had dementia for her last 10 years after an operation....[as did] my grandmother [...] I just had a hip replacement 2 months ago [...]. The odds are I may well end up with dementia...which I don't want to but wouldn't be surprising if I did'. P11, Female, 68 When explaining her family's experience with informal care and mentioning she did not have children, she expressed concerns about later life care: 'The health and social care system depends on family members, not just for doing things but for advocating... then it's quite scary to think you don't have access to that yourself.' P8, Male, 33, also mentioned that his choice of not having children, plus not wanting to be a burden to his sister in later life, meant that his meaningful connections in later life would not be family members but that he expected the health system to offer care when required.

4 Discussion

This paper aimed to explore the conceptualisation of a residential BE for ageing-in-place using coproduction. Thematic analysis revealed three relevant areas: **1) Micro-scale preferences:** Findings align with previous research about the challenges for social relations, connectivity and place attachment (Barken 2021; Seamon 2021) in rural settings (Porteus 2018; Skinner 2021). Results regarding dwelling layout show some similarities to the features suggested by Pickles (1998) (i.e. wide doors and wheelchair-accessible circulation spaces); however other features (i.e. having 3–4 bedrooms) differ from literature where typical layouts include 1–2 bedrooms to facilitate property management and reduce energy costs, assuming that downsizing (Park et al. 2016) is bound with ageing-in-place. These differences could be associated with perceived property value (Zahirovich-Herbert and Gibler 2014). The authors suggest further research with participants with a new-build strategy for ageing-in-place. **2) Technologies and interactions,** which highlight the need to use technologies to enhance human-human connections (Briede-Westermeyer et al. 2020) and deliver services that promote these interactions instead of attempting to substitute for them (Sikali 2020), as well as the ability to select and implement according to their personal preferences (De Vriendt et al. 2013; Cornelis et al. 2019; Dias et al. 2019). **3) Transferring informal care experiences into strategies**

to age in place, which showed how caring for loved ones can inform housing choices in later life (i.e. choosing to move to a single-storey property) and lifestyle influenced (i.e. keeping track of physical activity). Another discovery was that participants did not describe their informal care experiences as a burden; however, when foreseeing scenarios for their future, they did; this could be attributed to their particular experiences of care provision, as suggested by (Collins and Kishita 2020; Rodríguez-González et al. 2021); so collecting and analysing data regarding the type and length of care and conditions in further research should be considered and included when using the Positive Interdependence Framework.

This paper can be used as a reference for further studies incorporating coproduction to involve end-users in the development phase of the Double Diamond Method (Clune and Lockrey 2014; Cahya et al. 2018; Fleury and Richir 2021; Miolo de Oliveira et al. 2023).

5 Conclusions

The present study provides an initial framework to integrate informal care provision to shape a concept for housing. The outcomes show how coproduction can be used to conceptualise and develop housing alternatives for ageing in place, including insights about preferred home configurations and how physical and digital interactions with products available in the market could improve quality of life as people age. This paper can be used as a reference for designing a more inclusive BE, home adaptations, smart home technologies and assisted-living technologies. The Positive Interdependence Framework can help conceptualise the BE and/or its interventions (i.e. home adaptations, assisted living technologies) and support people in maintaining functional ability, specifically by mapping advanced activities of daily living. By using this approach, independent living strategies could evolve to support positive interdependence networks of people using digital and physical spaces involving and enhancing self-care and informal and formal care services.

Acknowledgements. The authors would like to extend thanks to all participants; to Innovate UK for funding this project (KTP011578) and the Participatory Research Fund 2022 from Teesside University.

References

Alzheimer's Society: Alzheimer's Society co-creation guide (2022). www.alzheimers.org.uk/sites/default/files/2022-05/alzheimerssociety-co-creation-guide.pdf. Accessed 1 July 2022

Anderten-Loeb, P.: Independent living scales (ILS) (2003). www.pearsonclinical.co.uk/store/ukassessments/en/Store/Professional-Assessments/Cognition-%26-Neuro/Independent-Living Scales/p/P100009018.html and https://libguides.lib.umanitoba.ca/c.php?g=297419&p=1993864. Accessed 17 Jan 2023

Ball, M., Nanda, A.: Household attributes and the future demand for retirement housing. Int. J. Hous. Mark. Anal. **6**(1), 45–62 (2013). https://doi.org/10.1108/17538271311306002

Barken, R.: The temporal aspects of mobility intentions: Older people's reflections on present and future support arrangements. Ageing Soc. **41**(3), 561–581 (2021). https://doi.org/10.1017/S0144686X19001223

Braun, V., Clarke, V.: Thematic Analysis: A Practical Guide. SAGE, Los Angeles (2022)

Briede-Westermeyer, J.C., Pacheco-Blanco, B., Luzardo-Briceño, M., Pérez-Villalobos, C.: Mobile phone use by the elderly: relationship between usability, social activity, and the environment. Sustainability **12**(7), 2690 (2020)

Brookfield, K., Scott, I., Tinker, A., Ward Thompson, C.: Perspectives on "novel" techniques for designing age-friendly homes and neighbourhoods with older adults. Int. J. Environ. Res. Public Health **17**(5), 1800 (2020). https://doi.org/10.3390/ijerph17051800

Cahya, R.A.D., Handayani, A.N., Wibawa, A.P.: Mobile Braille touch application for visually impaired people using double diamond approach. MATEC Web Conf. **197**, 15007 (2018). https://doi.org/10.1051/matecconf/201819715007

Carnemolla, P.: Ageing in place and the internet of things – how smart home technologies, the built environment and caregiving intersect. Vis. Eng. **6**(1), 1–16 (2018). https://doi.org/10.1186/s40327-018-0066-5

Caron, R.M.: Expansion of coproduction in health fare as a population health management approach. Popul. Health Manag. **22**(4), 286–288 (2019). https://doi.org/10.1089/pop.2018.0100

Clune, S.J., Lockrey, S.: Developing environmental sustainability strategies, the Double Diamond method of LCA and design thinking: a case study from aged care. J. Clean. Prod. **85**, 67–82 (2014). https://doi.org/10.1016/j.jclepro.2014.02.003

Collins, R.N., Kishita, N.: Prevalence of depression and burden among informal care-givers of people with dementia: a meta-analysis. Ageing Soc. **40**(11), 2355–2392 (2020). https://doi.org/10.1017/S0144686X19000527

Cornelis, E., Gorus, E., Van Schelvergem, N., De Vriendt, P.: The relationship between basic, instrumental, and advanced activities of daily living and executive functioning in geriatric patients with neurocognitive disorders. Int. J. Geriatr. Psychiatry **34**(6), 889–899 (2019). https://doi.org/10.1002/gps.5087

Crews, D.E.: Aging, frailty, and design of built environments. J. Physiol. Anthropol. **41**(1), 2 (2022). https://doi.org/10.1186/s40101-021-00274-w

De Vriendt, P., Gorus, E., Cornelis, E., Bautmans, I., Petrovic, M., et al.: The advanced activities of daily living: a tool allowing the evaluation of subtle functional decline in mild cognitive impairment. J. Nutr. Health Aging **17**(1), 64–71 (2013). https://doi.org/10.1007/s12603-012-0381-9

Design Council: Eleven lessons: Managing design in eleven global companies (2007). www.designcouncil.org.uk/our-work/skills-learning/resources/11-lessons-managing-design-global-brands/. Accessed 19 Jan 2023

Dias, E.N., da Silva, J.V., Pais-Ribeiro, J.L., Martins, T.: Validation of the advanced activities of daily living scale. Geriatr. Nurs. **40**(1), 7–12 (2019). https://doi.org/10.1016/j.gerinurse.2018.05.008

Enkvis, A., Ekstrom, H., Elmstahl, S.: Associations between functional ability and life satisfaction in the oldest old: results from the longitudinal population study good aging in Skåne. Clin. Interv. Aging **7**, 313–320 (2012). https://doi.org/10.2147/CIA.S33610

Fleury, S., Richir, S.: User-centered innovation methods: design thinking, double diamond, lean UX and time to concept. In: Fleury, S., Richir, S. (eds.) Immersive Technologies to Accelerate Innovation, pp. 67–81. Wiley, Hoboken (2021). https://doi.org/10.1002/9781119887294.ch4

Haux, R., Hein, A., Kolb, G., Künemund, H., Eichelberg, M., et al.: Information and communication technologies for promoting and sustaining quality of life, health and self-sufficiency in ageing societies – outcomes of the Lower Saxony research network design of environments for ageing (GAL). Inform. Health Soc. Care **39**(3–4), 166–187 (2014). https://doi.org/10.3109/17538157.2014.931849

Hu, R., et al.: Developing a smart home solution based on personalized intelligent interior units to promote activity and customized healthcare for aging society. J. Popul. Ageing **13**(2), 257–280 (2020). https://doi.org/10.1007/s12062-020-09267-6

Lopez, J., Textor, C., Braxton Hicks, W., Pryor, M., Collins McLaughlin, A., et al.: An aging-focused heuristic evaluation of home automation controls. In: Proceedings of the Human Factors and Ergonomics Society Annual Meeting, vol. 63, no. 1, pp. 6–10 (2019). https://doi.org/10.1177/1071181319631446

Lu, S., Liu, Y., Guo, Y., Ho, H.C., Chui, H.K., et al.: Neighborhood environment, intrinsic capacity and 4-year late-life functional ability trajectories. Innov. Aging **5**(Suppl. 1), 468–468 (2021). https://doi.org/10.1093/geroni/igab046.1808

Miolo de Oliveira, C., et al.: EU4ALL: for a strategic communication on the democratic values highlighted by the european parliament to groups vulnerable to information. In: Springer Series in Design and Innovation, vol. 27, pp. 345–356. Springer, Cham (2023). https://doi.org/10.1007/978-3-031-20364-0_30

Moreno-Agostino, D., Prina, M., Chua, K.-C., Jotheeswaran, A.T., Sadana, R., et al.: Measuring functional ability in healthy ageing: a nationwide cross-sectional survey in the Philippine older population. BMJ Open **11**(10), e050827–e050827 (2021). https://doi.org/10.1136/bmjopen-2021-050827

Park, A., Ziegler, F., Wigglesworth, S.: Designing with downsizers (2016). www.housinglin.org.uk/_assets/DWELL_DesigningWithDownsizers.pdf. Accessed 18 Jan 2023

Peace, S.: The Environments of Ageing: Space, Place and Materiality. Bristol University Press, Bristol (2022)

Peek, S.T.M., Wouters, E.J.M., van Hoof, J., Luijkx, K.G., Boeije, H.R., et al.: Factors influencing acceptance of technology for aging in place: a systematic review. Int. J. Med. Informatics **83**(4), 235–248 (2014). https://doi.org/10.1016/j.ijmedinf.2014.01.004

Pickles, J.: Housing for varying needs: A design guide. The Stationery Office, Edinburgh, UK (1998)

Pirinen, A.: Housing concepts for and by the elderly: from subjects of design to a design resource. J. Hous. Elder. **30**(4), 412–429 (2016). https://doi.org/10.1080/02763893.2016.1224792

Porteus, J.: Rural housing for an ageing population: preserving independence. HAPPI 4. The rural HAPPI inquiry (2018). www.housinglin.org.uk/_assets/Resources/Housing/Support_materials/Reports/HAPPI-4-Rural-Housing-for-an-Ageing-Population.pdf. Accessed 18 Jan 2023

Rodríguez-González, A.M., Rodríguez-Míguez, E., Claveria, A.: Determinants of caregiving burden among informal caregivers of adult care recipients with chronic illness. J. Clin. Nurs. **30**(9–10), 1335–1346 (2021). https://doi.org/10.1111/jocn.15683

Rogers, W.A., Kadylak, T., Bayles, M.A.: Maximizing the benefits of participatory design for human–robot interaction research with older adults. Hum. Factors **64**(3), 441–450 (2021). https://doi.org/10.1177/00187208211037465

Seamon, D.: Place attachment and phenomenology: the dynamic complexity of place. In: Manzo, L.C., Devine-Wright, P. (eds.) Place Attachment: Advances in Theory, Methods and Research, 2nd edn, pp. 29–44. Rougledge, New York (2021)

Sikali, K.: The dangers of social distancing: how COVID-19 can reshape our social experience. J. Community Psychol. **48**(8), 2435–2438 (2020). https://doi.org/10.1002/jcop.22430

Skinner, M., Winterton, R., Walsh, K. (eds.): Rural Gerontology. Towards Critical Perspectives on Rural Ageing. Routledge, New York (2021)

WHO: Ageing: Healthy ageing and functional ability, 26 October 2020 (2020a). www.who.int/news-room/questions-and-answers/item/healthy-ageing-and-functional-ability. Accessed 15 Nov 2020

WHO. Decade of healthy ageing: Baseline report. World Health Organization, Geneva, Switzerland (2020b)

WHO: WHO clinical consortium on healthy ageing, 2021. Departmental News (2022). www.who.int/news/item/02-11-2021-who-clinical-consortium-on-healthy-ageing-meeting-2021. Accessed 18 Jan 2023

Wilton, C.: Coproduction and partnership with people and communities. BMJ Leader **5**(2), 79–82 (2021). https://doi.org/10.1136/leader-2020-000424

Zahirovich-Herbert, V., Gibler, K.M.: The effect of new residential construction on housing prices. J. Hous. Econ. **26**, 1–18 (2014). https://doi.org/10.1016/j.jhe.2014.06.003

Examining Inclusion of Women in Public Spaces: A Review of Tools and Methods

A. Isha(✉) and G. Raheja

Department of Architecture and Planning, Indian Institute of Technology-Roorkee, Roorkee, India

{a_isha,gaurav.raheja}@ar.iitr.ac.in

Abstract. Public spaces provide people with a platform for political, social, and cultural activities, greatly enhancing the quality of life for all. But not everyone benefits equally from these spaces. Across the world, women are seen to use public spaces differently from men for various reasons. This study is an attempt to understand the ongoing application of tools and methods that are used to study how women perceive and use spaces differently from men. It is a narrative-based review of the published papers on women-centric studies in urban public spaces to highlight the tools and methods being used. A total of 28 papers were reviewed, which included studies of countries from both Global North and Global South. The results show that the most common choice of method is direct observations, followed by surveys, in-depth interviews, mapping, and photography. The paper demonstrates that the most effective way to collect women-centric data is to apply mixed methods. It also reveals some interesting and unconventional methods, like crowdmapping and social media, that are being used to collect large-scale data without going into the field. The paper concludes with the need for an intersectional approach in selecting the tools and methods to identify how women's sociocultural identities create varying degrees of exclusion from public spaces, and how these can be measured.

1 Introduction

Public spaces are an essential part of our urban infrastructure. They provide social and health-related functions, and they play a key role in improving the quality of life and well-being of city dwellers (Polko and Kimic 2022). However, public spaces have never been completely inclusive for all (Mehta 2014). Globally, women represent 49.58% of the world population (StatisticsTimes.com 2021). They not only use and perceive cities in a different way from men, but at the same time, they also use public places in diverse ways. (UN Women n.d.). In most cases, their travel pattern is impacted by their greater dependency on public transportation. Most women visit a public space with a sense of purpose rather than the intention of simply loitering around for recreation (Ollivier et al. 2022). Preference for any public space is influenced by the availability of amenities like safe and clean public toilets for women, with childcare facilities, adequate illumination after dark, last-mile connectivity, presence/absence of people in that space, etc. Gender differences exist in how people use urban open spaces, and the suggested reasons

J. Goodman-Deane et al. (Eds.): CWUAAT 2023, *Design for Sustainable Inclusion*, pp. 83–91, 2023.
https://doi.org/10.1007/978-3-031-28528-8_9

are mostly related to preferences, cultural norms, and individual choices. Additionally, women are more often observed to be in small family groups, their activities reflecting traditional gender and family roles (Huang and Napawan 2021). In most cultures, women are more likely to be victims of violence; as a result, even if women have access to a particular public space, they may not feel safe using it (Valentine 1990). It is worth mentioning that throughout the paper, gender is seen as nonbinary and derived from an individual's personal identity based on a spectrum of biological, mental and emotional traits. This paper only examines cases of inclusion for people in public spaces who identify themselves as women.

The term 'gender inclusive' denotes an approach that has an inclusive view of all genders, and their intersectionality like race, ethnicity, income, class, age, and ability, to make sure everyone's opinion is considered, and the projects are designed, delivered and evaluated to promote gender equity (Terraza et al. 2020). The inclusion of women in public space studies is a complex domain owing to its qualitative and subjective character. A woman's experience in a public space differs depending on the public space and it may vary from one woman to another. Women's experience of public space is related to its design. Gender-inclusive design strategies require a comprehensive understanding of gendered differences in the experience of space in specific contexts (Huang and Napawan 2021).

There are tangible and intangible elements that have an impact on the use of public space. While assessing the inclusion of women in public spaces, it is easier to observe the tangible elements but very often the intangible ones get overlooked. So the selection of the right tool and methodology is important to analyse a public space with a holistic approach.

The Sustainable Development Goal 11.7 aims to give universal access to safe, inclusive, and accessible green and other public places, especially for women, children, the elderly and persons with disabilities (The Global Goals 2022). This further highlights the importance creating inclusive public spaces for everyone. Gender disaggregated data in any field are useful to identify gaps, issues and challenges and make it easier to address them with suitable solutions. Similarly, we need innovative tools and methods to collect quantitative and qualitative data from the field to understand and analyse the preferences and perceptions of women about those spaces and suggest ways to make them inclusive to women. This study is a desk-based narrative review of research papers particularly looking for a focused method to study women and their everyday experiences in public spaces. Its primary aim was to identify existing tools and methods to map the perceptions of women in public spaces. The paper brings together in one place different tools being used in this domain, for easy reference, and discusses their role in various contexts for similar future studies.

2 Material and Methods

The review was done using the online tool Google Scholar and covered 28 published papers. The literature was assessed based on specific keywords (gender, public spaces, women, inclusion, urban park, built environment, and social inclusion) in published works related to studies regarding women and their presence in urban spaces. The search was further streamlined by considering titles and abstracts for the term "Gender", and "public space". These directly dealt with women's safety and mobility were filtered out to those relevant to the chosen subject of women and their everyday use of public spaces. Only those papers were included in the review that utilised unique tools and had a strong method-based approach to evaluate the use and perception of public spaces by women. The papers reviewed were selected to cover the maximum from various countries from Global North and Global South. Only the studies conducted in the past two decades were selected from the existing literature to get the most up to date tools and methods. Those that reused tools and methodologies were excluded. It is to be noted this search method is limited and it may have underrepresented studies in similar fields. The review concentrates primarily on the methods of data collection. The data analysis part of the studies is beyond its scope.

3 Results

3.1 Overview of Reviewed Papers on Inclusion of Women in Public Spaces

The 28 papers examined in this review mostly come from the United States of America, the United Kingdom, Poland, Turkey, Iran, Iraq, Saudi Arabia, Pakistan, India, Bangladesh, China, Indonesia, Malaysia, New Zealand, and Australia. It should be noted that most studies have come from various countries in the Global South. The maximum number of papers came from India followed by Iran. The review shows that there is a significantly higher concentration of method-based studies on women and public spaces after the 2010s (Refer to Fig. 1).

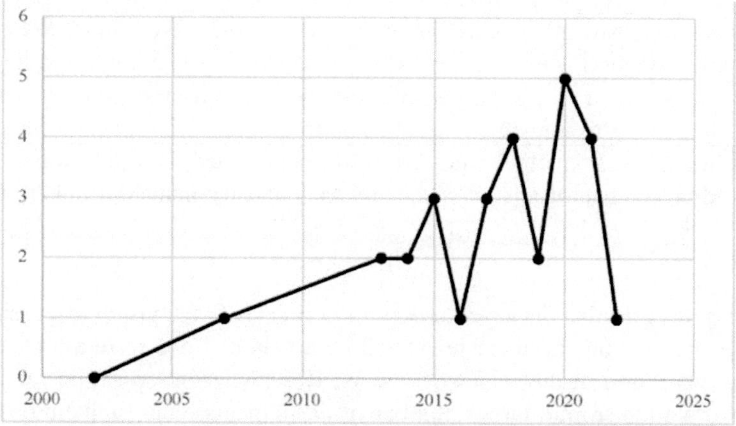

Fig. 1. Number of papers in the past two decades (2002–2022)

3.2 Tools and Methods Used in Reviewed Papers

The papers were reviewed according to their use of tools and methods that probe into the everyday lived experiences of women in public spaces. Most studies used a mixed-methods approach to study gender in public spaces. However, a few studies also came to light that had used new ways of data collection. The tools and methods extracted from the reviewed literature were classified on the basis of their use, the procedure of implementation and the medium under the following categories (Refer to Fig. 2): Observational Tools, Participatory Tools, Technology-Based Tools and Online-Based Tools.

1. Observational Tools. Mapping, counting, tracing, sketches and field notes were used to conduct studies of the papers reviewed. Mapping was used to record the activities of women– whether standing or sitting, either present in a group or alone, etc. (Gupta and Wadwekar 2018; Natu 2018; Mahadevia and Lathia 2019; Jabeen 2020; Zhang et al. 2022). Counting was extensively used to get the number of users in a particular space, their gender, counting items such as sitting spaces, streetlights, number of CCTV cameras, etc. Counting was also used to generate quantitative data. Tracing was used to understand the walking patterns of men and women on sidewalks in a study from Riyadh (Almahmood et al. 2017b). Tracing gave good details about women's circulation when used for neighbourhood-scaled studies. All of the above methods were used as a part of safety audits conducted in a case from New Delhi (Viswanath and Mehrotra 2007).

Photographs, videos and time-lapse recordings provided excellent information on women's behaviour in public spaces. Photography was the most popular method in most studies (Bagheri 2014a; Peimani and Kamalipour 2016; Almahmood et al. 2017b; Anwar et al. 2018; Gupta and Wadwekar 2018; Jabeen 2020; Lin et al. 2020; Roy and Bailey 2021). It was combined with other tools that gave very rich and deep insights regarding women's experiences and needs in public spaces. In gender inclusion studies, the photographs help to capture the emotions and behaviour of women that otherwise don't get recorded on paper. In the case of a study in Tehran, women showed specific facial expressions and other body languages when they talked about their feelings in public spaces (Bagheri 2014b). Such expressions could not be translated into words. In a study from Mumbai and Singapore, time-lapse video footage was shot across three locations in each city to study the loitering patterns of men and women in parks, roads and other public spaces. In another study from Mumbai, video recording was used to study the use and preference of leisure spaces by men and women (Mahimkar and Gokhale 2015).

2. Participatory Tools. Surveys, in-depth interviews, focus group discussions, participants' workshops, on-site interviews, and photo simulations were a few participatory tools that were used in the reviewed papers. Researchers used surveys extensively to gather information from a larger number of participants. One such study to look into gender disparities in parks was carried out in Poland by using a survey method. (Polko and Kimic 2022). The survey was able to take inputs regarding the spatial features and physical attributes of parks but the personal experiences remained undiscussed. However, the most interesting and relevant information regarding women's perception of using public spaces came from in-depth interviews (Johnson and Miles 2014; Tuncer 2015; Anwar et al. 2018; Natu 2018) and focus group discussions (Viswanath and Mehrotra 2007; Jabeen 2020; Roy and Bailey 2021; Zhang et al. 2022). During these interactions, women were observed to be candid about their experiences and wishes in public

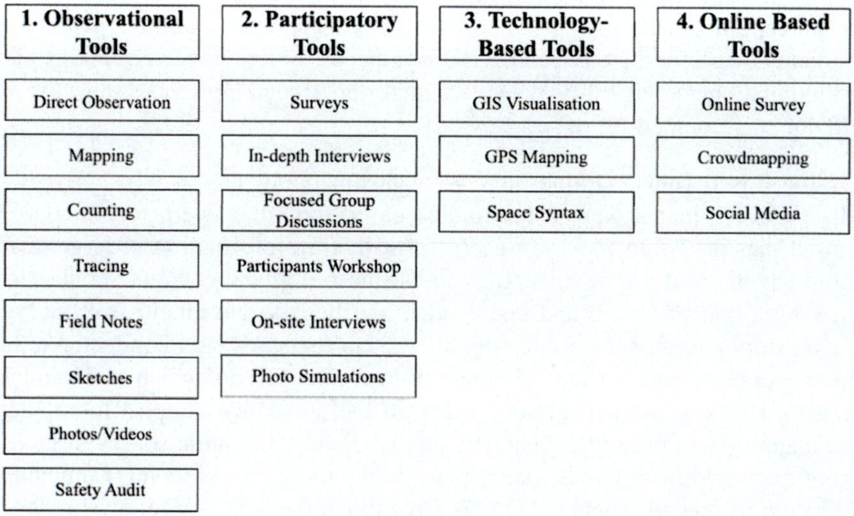

Fig. 2. Classification of reviewed tools and methods

spaces. They also explained why they chose to stay longer at a given location and their reason for choosing a different route from others. During such discussions, they even expressed their expectations of how public spaces could cater to their needs in a better way. In a case from Bhopal (Gupta and Wadwekar 2018) where there was a shortage of time to conduct in-depth interviews, on-site interviews were seen to be helpful. One study from London (Navarrete-Hernandez et al. 2021) used on-site photo-simulations to record participants' perception of safety as per gender by showing them the existing and proposed designs for certain parts of public space. In a study conducted in Riyadh, a participants' workshop was conducted after a GPS tracking exercise to understand the influence of sociocultural factors in the movement patterns of men and women using a street (Almahmood et al. 2017a).

3. Technology-Based Tools. Global Positioning System (GPS) tracking, Geographic Information System (GIS) visualisation and Space Syntax are relatively newer tools used for gendered data collection that have been in increasing use since the 2010s. They are useful to map the user movement patterns across built environments over a large scale and help to analyse why certain movement corridors are preferred over others. In one of the studies, GPS tracking was used to analyse the difference in walking patterns between men and women in a street in Riyadh (Almahmood et al. 2017a). In a study conducted in Tehran, Qualitative GIS (QGIS) visualisation was employed, linking images, sketches, and other narratives to the GIS data. As per the researcher, QGIS helped to manage large data sets, and understand the spatial use of social systems but they also mentioned that data exclusion occurred while trying to translate the narratives of women into the software (Bagheri 2014b). Space Syntax was used to conduct visibility analysis, map the connectivity of street layout and trace the pedestrian movement in a study from Turkey (Güney 2014). The study revealed the great divide between the use of streets by men and women at a city scale. It was also used in a case from Indonesia and Malaysia to study perception of safety by men and women by mapping the walking patterns around the street network for the city centres (Hidayati et al. 2020). However, smaller elements like illumination and visibility, which are important in making spaces safer for women, could not be accounted for in space syntax.

4. Online-Based Tools. Online surveys, crowdmapping and newly emerged social media posts/comments were a few online-based tools that were identified from the reviewed papers. These tools are useful in gathering information in large quantaties without having to in the field. A study in Northern England conducted online surveys containing a mix of closed and open-ended questions to get quantitative and qualitative data from participants (Roberts et al. 2022). The study was conducted to analyse the perceptions of safety and experiences of interpersonal violence by men and women during their time as students, both on and off the university campus. In another case, crowdmapping was used to collect data on experiences in public spaces from different parts of the world from female participants in the five cities of Delhi, Kampala, Lima, Madrid and Sydney (Tanner et al. 2020). Crowdmapping was helpful in situations when women were hesitant to share information with the researchers directly. This tool enabled them to stay anonymous and give their feedback and share experiences on this platform. In the study, crowdmapping allowed the female participants to submit feedback for specific geographic coordinates which led to high data accuracy. Real-time, spatial data set output could also be generated in a visual format for mass viewing. In another study, the social media platform of Facebook comments was reviewed to analyse the reactions of people over a feminist art piece installed in the city of Sulaimani, Iraq (Alizadeh et al. 2021). Data were collected from 10 popular Facebook pages that were chosen based on their popularity, number of followers and focus areas. Choosing this method gave the researchers information about natural responses from people but it failed to get demographic data (gender and location) of commentators.

4 Discussion

Any gender-space study would be incomplete without getting personal insights from women who have been using that particular space. It was found that every tool reviewed is also commonly used to assess human behaviour in general. However, the studies from the cases reviewed were conducted with a strong gender lens that greatly influenced the results to create women-centric data and present gaps in the existing built spaces that can be exclusionary to women. The review also draws the importance of mixing emerging technologies with participatory methods to gather a comprehensive understanding of women's needs, behaviour and perceptions of public spaces. Although online tools for data collection have been a revolutionary way of getting authentic information and responses from women, they fail to represent all sections of the target user groups due to lack of access to technology and the internet. The review saw the relevant role of the researcher in conducting gender studies. Women participants were hesitant to interact with a male researcher as seen in a case in Riyadh (Almahmood et al. 2017b). The major outcome of this paper is the compilation and classification of available tools and methodologies being used in similar studies across the world. This will benefit future studies on women and public spaces where researchers can refer to the list of tools mentioned. This study comes with certain limitations. The tools mentioned in the paper come from a few selected reviewed works. With the rising awareness of the need for inclusive cities, many other toolkits are also being used by urban practitioners that can be explored for conducting similar studies in the future.

5 Conclusion

The study is relevant for anyone, like architects, planners, urbanists, sociologists, urban local bodies, NGOs etc., who plans to examine public spaces with a gendered lens for women. Although each study has been done for a specific context, it has been realised that most tools and methods can be adapted to suit the requirements for future studies to break down the complex relationship between women and the built environment. It has been seen all four categories of tools; namely, Observational Tools, Participatory Tools, Technology-Based Tools and Online-Based Tools are effective to study cases with women and public spaces and they all come with their strengths and limitations. The review reveals the most commonly used tools are direct observations, surveys, in-depth interviews, mapping and photography. It shows that the most relevant information regarding the perception of women about public spaces comes from participatory tools like- in-depth interviews, focus group discussions, participant workshops and on-site interviews. Technology-based and online data collection does not have a large-scale application in places with inadequate technological and online support. A large part of the urban population in Global South countries still doesn't have access to digital and online facilities to conduct online studies. Social media posts can also be analysed and used to complement other tools used in public space and public life studies. This analysis can be combined with interviews and focus group discussions hosted on online meeting channels and online surveys, to validate existing observational data and survey results in cases where it is not possible to collect information from the field. The tools evaluated can be chosen based on the following factors: (1) the scale of implementation (city or neighbourhood scale); (2) the amount of time available for the study; and (3) the type of data needed—qualitative or quantitative. It is recommended to have an intersectional approach while selecting the tools and methods to identify how people's social identities overlap with the built environment and create varying degrees of exclusion from public spaces. The paper concludes with the importance of mixing the available four categories of tools in order to get the most holistic understanding of the existing issues faced by women in public spaces. The findings from the studies can be taken up with the local authorities to make suitable changes in the public spaces as suggested by the researchers.

References

Alizadeh, H., Kohlbacher, J., Mohammed-Amin, R.K., Raouf, T.L.: Gender inequalities and the effects of feminine artworks on public spaces: a dialogue. Soc. Inclus. **9**(4), 158–167 (2021)

Almahmood, M., Scharnhors, E., Carstensen, T.A., Jørgensen, G., Schulze, O.: Mapping the gendered city: investigating the socio-cultural influence on the practice of walking and the meaning of walkscapes among young Saudi adults in Riyadh. J. Urban Des. **22**(2), 229–248 (2017a)

Almahmoud, M., Schulze, O., Carstensen, T.A., Jørgensen, G.: The sidewalk as a contested space: women's negotiation of socio-spatial processes of exclusion in public urban space in Saudi Arabia; THE case of Al Tahlia Street. Plan. Pract. Res. **33**(2), 186–210 (2017b)
Anwar, N.H., Viqar, S., Mustafa, D.: Intersections of gender, mobility, and violence in urban Pakistan, In: Salahub, J.E., Gottsbacher, M., de Boer, J. (eds.) Social Theories of Urban Violence in the Global South, pp. 15–31. Routledge (2018)
Bagheri, N.: Mapping women in Tehran's public spaces: a geo-visualization perspective. Gend. Place Cult. **21**(10), 1285–1301 (2014a)
Bagheri, N.: What qualitative GIS maps tell and don't tell: Insights from mapping women in Tehran's public spaces. J. Cult. Geogr. **31**(2), 166–178 (2014b)
Gupta, N., Wadwekar, A.: Reclaiming gendered urban spaces by facilitating fearless movement of women in the fearful cities: Enabling inclusiveness by advocating spatial mediation. In: Proceedings of The IAFOR International Conference on the City 2018, Barcelona, Spain, 13–15 July (2018)
Güney, Y.İ.: Gender and urban space: an examination of a small Anatolian city. A|Z ITU J. Faculty Architect. **11**(2), 153–172 (2014)
Hidayati, I., Tan, W., Yamu, C.: How gender differences and perceptions of safety shape urban mobility in Southeast Asia. Transport. Res. Traffic Psychol. Behav. **73**, 155–173 (2020)
Huang, Y., Napawan, N.C.: "Separate but equal?" Understanding gender differences in urban park usage and its implications for gender-inclusive design. Landsc. J. **40**(1), 1–16 (2021)
Jabeen, H.: Gender-responsive urban planning and design of public open spaces for social and economic equity: challenges and opportunities in Dhaka City, Bangladesh. Glob. J. Hum.-Soc. Sci. E Econ. **20**(5), 1–14 (2020)
Johnson, A.M., Miles, R.: Toward more inclusive public spaces: Learning from the everyday experiences of Muslim Arab women in New York City. Environ. Plan A **46**(8), 1892–1907 (2014)
Lin, M., Bao, J., Dong, E.: Dancing in public spaces: an exploratory study on China's Grooving Grannies. Leis. Stud. **39**(4), 545–557 (2020)
Mahadevia, D., Lathia, S.: Women's safety and public spaces: lessons from the Sabarmati riverfront. India. Urban Planning **4**(2), 154–168 (2019)
Mahimkar, S., Gokhale, V.A.: Gender space dynamics of urban outdoor public leisure spaces, In: Proceedings of the International Conference on Changing Cities II: Spatial, Design, Lanscape & Socio-economic Dimensions. Porto Heli, Greece, 22–26 June, pp. 1649–1658 (2015)
Mehta, V.: Evaluating public space. J. Urban Des. **19**(1), 53–88 (2014)
Natu, A.: Gendered use of landscapes: a case of community open spaces (2018)
Navarrete-Hernandez, P., Vetro, A., Concha, P.: Building safer public spaces: exploring gender difference in the perception of safety in public space through urban design interventions. Landsc. Urban Plan. **214**, 104180 (2021)
Ollivier, G.P., Nikore, M., Lnu, S.N., Vasudevan, V., Shah, S.N., et al.: India-Toolkit for Enabling Gender Responsive Urban Mobility and Public Apaces: Volume 1–The'What-To-Do'Note for Policymakers. World Bank Group (2022).
Peimani, N., Kamalipour, H.: Where gender comes to the fore: Mapping gender mix in urban public spaces. Spaces Flows Int. J. Urban ExtraUrban Stud. **8**(1), 19–30 (2016)
Polko, P., Kimic, K.: Gender as a factor differentiating the perceptions of safety in urban parks. Ain Shams Eng. J. **13**(3), 101608 (2022)
Roberts, N., Donovan, C., Durey, M.: Gendered landscapes of safety: How women construct and navigate the urban landscape to avoid sexual violence. Criminol. Crim. Just. **22**(2), 287–303 (2022)
Roy, S., Bailey, A.: Safe in the City? Negotiating safety, public space and the male gaze in Kolkata. India. Cities **117**, 103321 (2021)

StatisticsTimes.com: Gender ratio in the World. 26 August 2021. https://statisticstimes.com/demographics/world-sex-ratio.php. Accessed 16 Jan 2023

Tanner, S., Kalms, N., Cull, H., Matthewson, G., Aisenberg, A.: Disruption and Design: Crowdmapping Young Women's Experience in Cities (2020)

Terraza, H., Orlando, M.B., Lakovits, C., Lopes, J.V., Kalashyan, A.: Handbook for gender-inclusive urban planning and design (2021). World Bank, Washington, DC . https://openknowledge.worldbank.org/handle/10986/33197 License: CC BY 3.0 IGO. Accessed 16 Jan 2023

The Global Goals: Goal 11: Sustainable cities and communities (2022). www.globalgoals.org/goals/11-sustainable-cities-and-communities/. Accessed 16 Jan 2023

Tuncer, S.: Going public: women's narratives of everyday gendered violence in modern Turkey. In: Proceedings of Papers of International Conference on Knowledge and Politics in Gender and Women's Studies (2015). https://doi.org/10.13140/RG.2.1.2111.6646

UN Women: Creating safe and empowering public spaces with women and girls. UN Women Headquarters (n.d.). www.unwomen.org/en/what-we-do/ending-violence-against-women/creating-safe-public-spaces. Accessed 16 Jan 2023

Valentine, G.: Women's fear and the design of public space. Built Environ. **1978**, 288–303 (1990)

Viswanath, K., Mehrotra, S.T.: 'Shall we go out?' Women's safety in public spaces in Delhi. Econ. Polit. Weekly, 1542–1548 (2007)

Zhang, Y., Chen, G., He, Y., Jiang, X., Xue, C.: Social interaction in public spaces and well-being among elderly women: towards age-friendly urban environments. Int. J. Environ. Res. Public Health **19**(2), 746 (2022)

Geography, Indoor Mobility and Activity Space: Informing Inclusive Healthcare Design

J. Sturge(✉)

Department of Design, Production and Management, Faculty of Engineering Technology, University of Twente, Enschede, The Netherlands
j.l.sturge@utwente.nl

Abstract. The design of the built environment of a healthcare setting impacts all users. As a result, it is becoming common practice to design and create healthcare environments that are inclusive of the needs of patients, staff and visitors. Translating how people interact with the environment into design requires a theoretical, methodological level of understanding the dynamics within the built environment. There are a number of social science concepts, methods and frameworks that demonstrate how individuals interact with environments. The aim of this chapter is to establish human geography, with a focus on indoor mobility and the concept of activity space, as a theoretical concept to explore the spatial and temporal movements of users of indoor healthcare environments. This new knowledge of spatial behavior can provide valuable insights into how indoor healthcare facilities work in practice and inform future design solutions based on users' needs.

1 Introduction

There is a growing interest in a research-based approach to healthcare architecture, particularly concerning sustainability, health and well-being and how the built environment shapes these outcomes. However, there remains a research-to-practice gap in informing the architectural design process in each area (Tvedebrink and Jelić 2021). Meanwhile, there is a movement to create socially sustainable buildings designed to benefit the users by understanding what they need from places and how they interact with built environments (Lami and Mecca 2021; Nenadović and Milošević 2022). Translating the dynamics of human interaction with built environments into design and architecture requires a theoretical and methodological understanding. Further, interdisciplinary collaborations from a range of research disciplines, with an interest in humans' interactions with buildings, are being encouraged to inform the design and construction of built environments that promote equality and inclusion of diverse users (Becerik-Gerber et al. 2022). Social scientists can bring a number of theories, methods and frameworks to understand how people interact with environments which can be then integrated into design. For instance, a design anthropological approach has been used to inform hospital design and architectural design processes (Pink et al. 2020). As a health geographer, I present the advantage of the discipline of human geography to explore the interaction between healthcare users and the environment. In addition, I propose extending the concept of activity space to explore mobility within healthcare environments so the design can improve usability and well-being.

J. Goodman-Deane et al. (Eds.): CWUAAT 2023, *Design for Sustainable Inclusion*, pp. 92–98, 2023.
https://doi.org/10.1007/978-3-031-28528-8_10

1.1 Geography and Healthcare Design

The discipline of geography has informed healthcare design for over thirty years. The most notable contribution is the early work of geographer Roger Ulrich who found patients recovered faster from surgery in hospital rooms with of nature views than those without (Ulrich 1984). This research is considered to be a foundation study of evidence-based design. Evidence-based design is an interdisciplinary, scientific process that translates research to improve health outcomes into the design of the built environment of healthcare facilities (Hamilton and Watkins 2009; Halawa et al. 2020). Related to evidence-based design is Ulrich (1991) stress reduction theory which identified design features, such as a controlled acoustic environment, glimpses of nature and daylight, that can improve patient health outcomes and reduce stress. This theory has been applied to the design of healthcare environments, including a Swedish psychiatric hospital. Studies confirmed that design features, such as floor layouts and natural views, reduced stress and aggression, and moreover improved patient and staff safety (Ulrich et al. 2018; Lundin 2021).

The geographic concept of therapeutic landscapes has also been applied to healthcare design. The therapeutic landscape concept, defined as 'how the healing process works itself out in places' (Gesler 1992: pp.743), was initially formulated to explore outdoor phenomena but later adapted as a framework to examine the relationship between place and health in indoor home spaces (Williams 2002). The concept of therapeutic landscapes has since been applied to several studies to explore places and indoor environments, including how the design of hospitals (e.g., inpatient psychiatric environments) can support or constrain health and relations between staff, patients and family (Curtis et al. 2007; Wood et al. 2015).

1.2 Indoor Mobility and Healthcare Design

Design features in healthcare settings can objectively positively impact patient outcomes (Elf et al. 2020). However, until recently, there was limited understanding of how healthcare design and architectural features come together and work in practice (i.e., are experienced by users and impact clinical practice) (Simonsen and Duff 2020a, b; Simonsen et al. 2022). One way to understand how features work in practice is to use a variety of research methods to explore patient, staff and visitor mobility within healthcare environments. For instance, Colley et al. (2018) used observation methods to explore how patients and staff used corridors in a rehabilitation environment as care and social space. This research suggests further research on how connective spaces (i.e., corridors) can be designed to contribute more deliberately to the positive experiences of all users. In another study in a rehabilitation setting, Annemans et al. (2022) used walking interviews and activity trackers to identify design features that hamper or motivate physical activity and social interaction. Kevdzija et al. (2022) mapped the wayfinding behaviour of persons with stroke in a rehabilitation environment through patient shadowing and a patient survey. This study identified patients' spatial preferences, including corridors and patient rooms. In another publication from the same study, the authors focus on the travel distances between spaces in rehabilitation clinics and how the distance and other barriers impacted patient mobility (Kevdzija and Marquardt 2022). Further, the configuration of hospital spaces has recently been investigated by adapting the social network analysis method. This study suggests that further research is needed to follow indoor mobility to capture routines and movement between different destinations (Sailer 2021).

Geographers have studied mobility and the factors that influence health and well-being over time and place for decades (Kwan and Schwanen 2016). Exploring the geographical dynamics of the movement of individuals and interaction within healthcare environments would complement indoor mobility research. This level of insight can be achieved through various data collection methods that capture mobility patterns and experiences of patients, staff and visitors. For instance, health geographers have used mixed methods (e.g., Global Positioning Systems, walking interviews and travel diaries) to identify features of the built environment that can support or hinder the well-being of specific groups such as older adults and persons with dementia (Meijering and Weitkamp 2016; Sturge et al. 2020). Most geographic mobility studies have focused on interactions with outdoor environments. However, a focus on movements within indoor environments is starting to emerge (Meijering 2019).

1.3 Activity Space and Healthcare Design

Activity space measurement of healthcare environments would complement indoor mobility studies by providing a spatial and temporal understanding of mobility. Activity space is a familiar concept in geography to explore mobility and spatial interactions with the environment. The concept was developed by Hägerstraand (1970) to explore daily travel behaviors based on space and time. Activity space can be defined as space outside the home that includes a subset of routinely and non-routinely visited locations and the corresponding travel routes which an individual uses over time (Golledge and Stimson 1998; Hirsch et al. 2016). Activity space is a valuable health research concept to explore environmental determinants of health and inequalities (Kestens et al. 2018), diet and physical activity (Zenk et al. 2011), accessibility of healthcare opportunities (Sherman et al. 2005) and social health of people with memory problems and dementia (Sturge et al. 2020). According to a recent systematic review, 47 articles have applied the concept of activity space to physical activity within to the environment (Smith et al. 2019). However, none of these studies identified indoor interactions, indicating a lack of research into them.

Conceptualising indoor mobility, including when and how long people are moving around specific locations, would provide a unique perspective on users' experiences within healthcare environments. For example, in dementia wards, there are a variety of places (i.e., activity nodes) and routes that residents take as a part of their daily routine (Fig. 1). The resident's activity space in such an environment can be measured and evaluated using different data collection methods. For instance, data from door sensors, step counts, walking interviews, participatory observations, and semi-structured interviews would provide an insight into how certain built environment features can enhance daily routines and well-being. Examining the time structure, such as the length of time spent at activity nodes or independently wandering between nodes, could inform the placement of assistive technologies or wayfinding features within the building. A focus on the size and structure of individual activity spaces may also optimise staffing by determining in advance the spaces that residents use and for which purpose. For instance, similar to the findings of Colley et al. (2018), Kevdzija et al. (2022) and Pink et al. (2020), it is likely that the corridors of dementia wards will be a common space (activity path) for interaction between residents and staff. Identifying how and when the residents choose to use the corridors could be beneficial for staff to plan for indirect, interactive care that meets residents' needs. Data can be analysed independently or further triangulated with other types of data, such as occupancy data, critical incidents or clinical assessment data. Combining data can provide a more in-depth understanding of the spatial configurations of the built environment that can impact mobility, patient outcomes and staff satisfaction. Further, comparing and layering spatial data using qualitative Geographic Information Systems (Cope and Elwood 2009) would provide an enhanced understanding of the day-to-day interactions of all users plus the routines and usability of the built environment. Such an overview of user interaction with the built environment would be valuable to inform operational decisions or identify features that need to be enhanced or renovated.

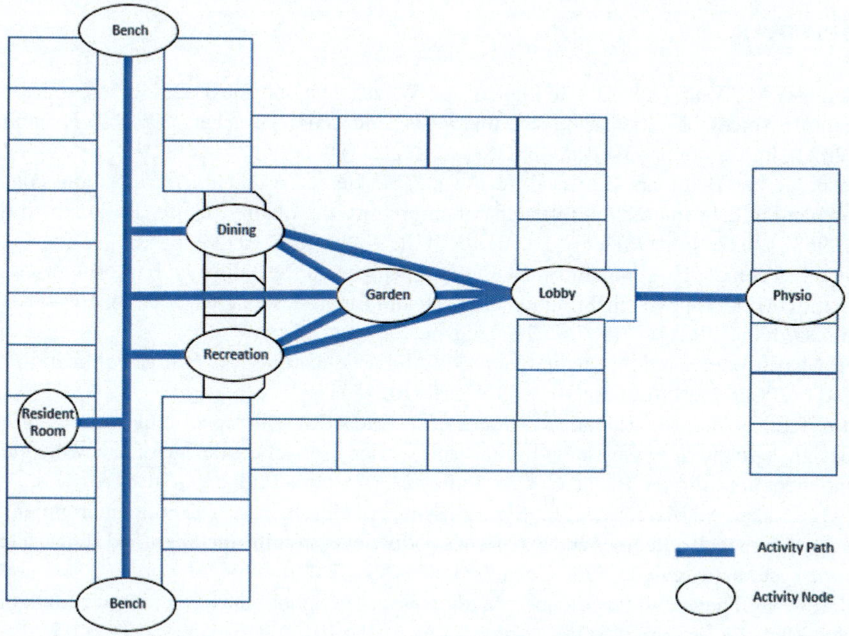

Fig. 1. An example of activity paths and nodes of a resident's activity space within a residential dementia care unit

2 Conclusions and Implications for Healthcare Design

To ensure that built healthcare environments are designed to support the diverse needs of patients, staff and visitors, researchers from various disciplines employ mixed methods to investigate the interaction between mobility and the built environment. Geographical methods, concepts and analyses would complement this research by providing insight into people and how they use the environments over time. In particular, adapting the concept of activity space to conceptualise indoor mobility would capture the complexity of the structure and distinguish the routines of patients, staff and visitors when they navigate and interact with the built environment on a daily basis. Exploring indoor time-space mobility with specific layouts, design features, and technology would enhance our ability to design environments that promote equality and inclusion. Additionally, a focus on activity space would provide a person-centred, theory-informed, comprehensive insight into the spatial configuration of indoor healthcare environments and how the built environment works in practice. This new direction for geography and activity space research promises to contribute to a growing knowledge base, including evidence-based design, human-building interaction research and post-occupancy evaluation. Translating this user-based knowledge across disciplines provides an opportunity to bridge a gap between theory and applied science resulting in the design of environments that improve the well-being and opportunities of all users.

References

Annemans, M., Van Dyck, D., Heylighen, A.: What affects physical activity in a rehabilitation centre? Voices of patients, nurses, therapists, and activity trackers. Disabil. Rehabil. 1–10 (2022). https://doi.org/10.1080/09638288.2022.2118873

Becerik-Gerber, B., Lucas, G., Aryal, A., Awada, M., Bergés, M., et al.: Ten questions concerning human-building interaction research for improving the quality of life. Build. Environ. **226**, 109681 (2022). https://doi.org/10.1016/J.BUILDENV.2022.109681

Colley, J., Zeeman, H., Kendall, E.: Everything happens in the hallways: Exploring user activity in the corridors at two rehabilitation units. Health Environ. Res. Des. J. **11**(2), 163–176 (2018). https://doi.org/10.1177/1937586717733149

Cope, M., Elwood, S. (eds.): Qualitative GIS. SAGE Publications Ltd. Online Publication January 2011 (2009). https://doi.org/10.4135/9780857024541

Curtis, S.E., Gesler, W., Fabian, K., Francis, S., Priebe, S.: Therapeutic landscapes in hospital design: a qualitative assessment by staff and service users of the design of a new mental health inpatient unit. Eviron. Plann. C. Gov. Policy (2007). https://doi.org/10.1068/c1312r

Elf, M., Anåker, A., Marcheschi, E., Sigurjónsson, Á., Ulrich, R.S.: The built environment and its impact on health outcomes and experiences of patients, significant others and staff—a protocol for a systematic review. Nurs. Open **7**(3), 895–899 (2020). https://doi.org/10.1002/nop2.452

Gesler, W.M.: Therapeutic landscapes: Medical issues in light of the new cultural geography. Soc. Sci. Med. **34**(7), 735–746 (1992). https://doi.org/10.1016/0277-9536(92)90360-3

Golledge, R.G., Stimson, R.J.: Spatial behaviour: a geographic perspective. Econ. Geogr. **74**(1), 83–85 (1998). https://doi.org/10.2307/144350

Halawa, F., Madathil, S.C., Gittler, A., Khasawneh, M.T.: Advancing evidence-based healthcare facility design: a systematic literature review. Health Care Manag. Sci. **23**(3), 453–480 (2020). https://doi.org/10.1007/s10729-020-09506-4

Hamilton, K., Watkins, D.H.: Evidence-Based Design for Multiple Building Types. Wiley, Hoboken (2009)

Hirsch, J.A., Winters, M., Ashe, M.C., Clarke, P.J., McKay, H.A.: Destinations that older adults experience within their GPS activity spaces: relation to objectively measured physical activity. Environ. Behav. **48**(1), 55–77 (2016). https://doi.org/10.1177/0013916515607312

Hägerstraand, T.: What about people in regional science? Pap. Reg. Sci. **24**(1), 7–24 (1970)

Kestens, Y., Thierry, B., Shareck, M., Steinmetz-Wood, M., Chaix, B.: Integrating activity spaces in health research: comparing the VERITAS activity space questionnaire with 7-day GPS tracking and prompted recall. Spat. Spatio-Temporal Epidemiol. **25**, 1–9 (2018). https://doi.org/10.1016/j.sste.2017.12.003

Kevdzija, M., Bozovic-Stamenovic, R., Marquardt, G.: Stroke patients' free-time activities and spatial preferences during inpatient recovery in rehabilitation centers. HERD: Health Environ. Res. Des. J. **15**(4), 96–113 (2022). https://doi.org/10.1177/19375867221113054

Kevdzija, M., Marquardt, G.: Impact of distance on stroke inpatients' mobility in rehabilitation clinics: a shadowing study. Build. Res. Inf. **50**(1–2), 74–88 (2022). https://doi.org/10.1080/09613218.2021.2001302

Kwan, M.-P., Schwanen, T.: Geographies of mobility. Ann. Am. Assoc. Geogr. 1–14 (2016). https://doi.org/10.1080/24694452.2015.1123067

Lami, I.M., Mecca, B.: Assessing social sustainability for achieving sustainable architecture. Sustainability **13**(1), 142 (2021). https://doi.org/10.3390/su13010142

Lundin, S.: Can healing architecture increase safety in the design of psychiatric wards? Health Environ. Res. Des. J. **14**(1), 106–117 (2021). https://doi.org/10.1177/1937586720971814

Meijering, L.: Towards meaningful mobility: a research agenda for movement within and between places in later life. Ageing Soc. **41**(4), 711–723 (2019). https://doi.org/10.1017/S0144686X19001296

Meijering, L., Weitkamp, G.: Numbers and narratives: developing a mixed-methods approach to understand mobility in later life. Soc. Sci. Med. **168**, 200–206 (2016). https://doi.org/10.1016/j.socscimed.2016.06.007

Nenadović, A., Milošević, J.: Creating sustainable buildings: Structural design based on the criterion of social benefits for building users. Sustainability **14**(4), 2133 (2022). https://doi.org/10.3390/su14042133

Pink, S., Duque, M., Sumartojo, S., Vaughan, L.: Making spaces for staff breaks: a design anthropology approach. Health Environ. Res. Des. J. **13**(2), 243–255 (2020). https://doi.org/10.1177/1937586719900954

Sailer, K.: Routine action networks: an architectural study of spatial layouts and performativity in outpatient clinics. Soc. Netw. (2021). https://doi.org/10.1016/J.SOCNET.2021.12.004

Sherman, J.E., Spencer, J., Preisser, J.S., Gesler, W.M., Arcury, T.A.: A suite of methods for representing activity space in a healthcare accessibility study. Int. J. Health Geogr. **4**(23) (2005). https://doi.org/10.1186/1476-072X-4-24

Simonsen, T., Sturge, J., Duff, C.: Healing architecture in healthcare: a scoping review. HERD: Health Environ. Res. Des. J. **15**(3), 315–328 (2022). https://doi.org/10.1177/19375867211072513

Simonsen, T., Thorben, P., Duff, C.: Healing architecture and psychiatric practice: (re)ordering work and space in an in-patient ward in Denmark. Sociol. Health Illn. **42**(2), 379–392 (2020). https://doi.org/10.1111/1467-9566.13011

Simonsen, T., Thorben, P., Duff, C.: Mutual visibility and interaction: Staff reactions to the 'healing architecture' of psychiatric inpatient wards in Denmark. BioSocieties **16**, 249–269 (2020). https://doi.org/10.1057/s41292-020-00195-4

Smith, L., Foley, L., Panter, J.: Activity spaces in studies of the environment and physical activity: a review and synthesis of implications for causality. Health Place **58**, 102113 (2019). https://doi.org/10.1016/j.healthplace.2019.04.003

Sturge, J., Klaassens, M., Lager, D., Weitkamp, G., Vegter, D., et al.: Using the concept of activity space to understand the social health of older adults living with memory problems and dementia at home. Soc. Sci. Med. **288**, 113208 (2020). https://doi.org/10.1016/j.socscimed.2020.113208

Tvedebrink, T.D.O., Jelić, A.: From research to practice: is rethinking architectural education the remedy? Health Environ. Res. Des. J. **14**(1), 71–86 (2021). https://doi.org/10.1177/1937586720953529

Ulrich, R.S.: View through a window may influence recovery from surgery. Science **224**(4647), 420–421 (1984). https://doi.org/10.1126/science.6143402

Ulrich, R.S.: Effects of interior design on wellness: theory and recent scientific research. J. Health Care Inter. Des. **3**, 97–109 (1991)

Ulrich, R.S., Bogren, L., Gardiner, S.K., Lundin, S.: Psychiatric ward design can reduce aggressive behavior. J. Environ. Psychol. **57**, 53–66 (2018). https://doi.org/10.1016/j.jenvp.2018.05.002

Williams, A.: Changing geographies of care: employing the concept of therapeutic landscapes as a framework in examining home space. Soc. Sci. Med. **55**(1), 141–154 (2002). https://doi.org/10.1016/S0277-9536(01)00209-X

Wood, V.J., Gesler, W., Curtis, S.E., Spencer, I.H., Close, H.J., et al.: Therapeutic landscapes and the importance of nostalgia, solastalgia, salvage and abandonment for psychiatric hospital design. Health Place **33**, 83–89 (2015). https://doi.org/10.1016/j.healthplace.2015.02.010

Zenk, S.N., Schulz, A.J., Matthews, S.A., Odoms-Young, A., Wilbur, J., et al.: Activity space environment and dietary and physical activity behaviors: a pilot study. Health Place **17**(5), 1150–1161 (2011). https://doi.org/10.1016/J.HEALTHPLACE.2011.05.001

Healthcare

Public Policies and Inclusive Design: Analysis of a Brazilian Oncology Service

P. R. Bohn[1(✉)], E. Konig[1], L. M. Tonetto[2], and C. S. Libânio[1]

[1] Federal University of Health Sciences of Porto Alegre (UFCSPA), Porto Alegre, Brazil
patriciabohn07@gmail.com, {emanuele,claudiasl}@ufcspa.edu.br
[2] National Council for Scientific and Technological Development, Brasília, Brazil
leandro.tonetto@pq.cnpq.br

Abstract. Legislation, standards, and policies (LSP) are essential means to promote and regulate accessibility, i.e., inclusion of vulnerable groups in society, especially concerning healthcare services. This study aims to analyse how public policies relate to inclusive design (ID) by promoting accessibility and inclusion in a Brazilian oncology service. We performed an exploratory case study; we analysed the Brazilian LSP, conducted in-depth interviews, and carried out non-participant observation. The results were analysed through thematic analysis and categorised into the following themes: service access, architectural and environmental aspects, information and communication aspects, technological aspects, and attitudinal aspects. In summary, we found that despite Brazilian LSP promoting the inclusion of patients in health services, the observations and interviews showed that there are many gaps in the area of inclusion and accessibility. The results highlight the complexity of working within political and regulatory contexts when implementing and managing such public policies. Then implementation and management should not be ignored, as they ensure the effectiveness of ID; long-term and ongoing actions at different levels of policy, implementation, and management must be taken to provide a more comprehensive understanding of ID in healthcare services.

1 Introduction

According to the United Nations (2016), people worldwide face exclusion, lack of access to opportunities, and comfortable participation in society. In the context of services, people have been excluded across human history, not being assisted adequately and fairly (Fisk et al. 2018). Given this scenario, inclusive design (ID) seeks to consider user needs in order to make it possible for people, especially those with reduced functional capabilities, to use products, services and environments without difficulties (Persad et al. 2007).

Healthcare services may impose certain limitations on patients, who need to be capable enough to access and use them in an integrated, fair, and satisfactory manner (Liu et al. 2021). Thus, the ID principles are important to design services that meet patients' needs, enabling proper access and use (Liu et al. 2021). This work is focused on the experience of patients undergoing cancer treatment. These patients have a series

J. Goodman-Deane et al. (Eds.): CWUAAT 2023, *Design for Sustainable Inclusion*, pp. 101–107, 2023.
https://doi.org/10.1007/978-3-031-28528-8_11

of vulnerabilities, disabilities, and limitations caused by the disease itself, the treatment, or pre-existing conditions exacerbated by the disease/treatment (Naughton and Weaver 2014). Since ID embraces the needs of persons with disabilities, older people, individuals from diverse social and cultural backgrounds, and others in vulnerable conditions (GDIH 2022), we highlight the importance of using the lens and principles of ID for oncological patients and services.

Patients diagnosed with cancer should be treated in a holistic, multidisciplinary, and humanised way (Costa et al. 2003). Although the importance of humanised assistance is noted and recognised, not all services and professionals recognise and apply it. So, several barriers undermine fair and equitable access to health services, such as architectural (Geraldo and Andrade 2022), technological (Carayon et al. 2019), attitudinal (Hashemi et al. 2022), and informational/communicational aspects (Geraldo and Andrade 2022).

Brazil has a set of laws, standards, and policies (LSP) that govern accessibility, inclusion, and health, seeking to guarantee the right of access of all people. Accessibility in this study is defined as the possibility and condition of needing the safe and autonomous use of spaces, furniture, urban equipment, buildings, transport, information, and communication, including their systems and technologies (Brazil 2015). The formulation of these public policies seeks to promote social sustainability for cancer patients and their families, based on the fulfillment of their needs. Despite all efforts, inclusive actions in healthcare end up not being promoted, causing marginalisation (Garnelo et al. 2020). Therefore, this study aims to analyse how public policies relate to ID by promoting accessibility and inclusion in a Brazilian oncology service.

2 Method

A case study was carried out in an outpatient oncological service of a Brazilian hospital. For data collection, initially, we analysed the LSP existing in Brazil that aim to promote the accessibility and inclusion of people in health services, especially oncology services. Afterwards, we identified and collected these documents in the digital databases of the bodies responsible for regulating and managing this issue in Brazil. Then we conducted in-depth interviews and carried out non-participant observations. The study was approved by the UFCSPA's Institutional Review Board (Bohn et al. 2022).

The service under analysis assists around 1 200 unified health systems (SUS) patients. This service provides consultations, medium-complexity examination, surgical treatment, and chemotherapy. We observed external and internal spaces of the outpatient clinics for surgical consultations, the environments of medical examinations and the chemotherapy space. The surgical consultation offices have a reception, two medical rooms, a bathroom, an administration room, and a dining area for employees. To access this environment, the patients need to go up a slope, and there is an access ramp to the door. The chemotherapy room is accessed right behind the offices, through a flat corridor outside the building. It has a reception, two medical offices, an oncology pharmacy, and the chemotherapy application room, which has a stretcher and 8 hospital sofa-chairs. This room is shared and patients do not have individual spaces to receive treatment. There are two spaces for carrying out clinical examinations: one on the ground floor and the other one on a different floor in the same building. One of them is intended for oncological referrals from SUS. The other is intended for hospitalisation, emergency care, and for patients contaminated by COVID-19. These environments have a reception, a screening room, the examination rooms (breast ultrasound, mammography, tomography, magnetic resonance, general ultrasound and X-ray), and administrative rooms.

We carried out in-depth interviews with three health professionals and six users of the outpatient oncological service. The interviews followed a semi-structured script that explored the patient's experience. They were recorded and transcribed, registered in field diaries by the two researchers separately, and the data was tabulated for further analysis.

Regarding observation, this was indirect and took place for eight days, in different shifts, according to the availability of the service and prior arrangement with the staff. In the observation process, the researchers did not interact with the patients, moving through the service spaces in non-participatory ways capturing information. Field diaries guided the process and the following aspects were observed: the identity of the place and people involved at the time of observation; patients' interactions with the service; impressions; and what needs further clarification.

For data analysis, thematic analysis was performed (Braun and Clarke 2014).

3 Results

Data was classified into five categories, considering aspects related to: access to the service, architecture and environment, information and communication, technology, and attitudes.

Concerning **service access**, data refer to universal and equitable approaches to adequate treatment. We verified that the Brazilian legislation presents guidelines regarding the form of access to health services and the deadline for patient access to treatment upon diagnosis of cancer (Brazil 2021). SUS seeks to ensure universal, equitable, and integral access for all people in Brazilian territory (Brazil 1988; Brazil 1990). The form of access considers the degree of complexity. In terms of time, access to services generally depends on the schedule of the respective services demanded. However, for patients diagnosed with cancer, there is a legal provision for access within 60 days to start treatment at a reference oncology service (Brazil 2021). Although we identified a robust conception of LSP, the form of access is not as effective as public policies predict and is poorly understood by a significant number of cancer patients, impending their access to the service, and confronting the principles of SUS. Cancer patients report access to specialist consultation within the established deadline, but the effective start of treatment takes longer than the established deadline, as do examinations. So, most of the so-called access routes were not equally available to all patients. High demand is one of the reasons.

Regarding **architectural and environmental aspects**, data refer to any issue in spaces and buildings that may promote or hinder a person from accessing, moving in and making full use of them safely. Brazilian LSP pay special attention to these aspects, ensuring that people with disabilities and reduced mobility can integrally access physical spaces, with necessary adaptations (Brazil 2002a, 2002b; (Brazil 2011). Especially NBR 9050 brings detailed specifications for built spaces, highlighting, for example, aspects of signage (visual, sound, and tactile) capable of finding for any user any place they want to reach and access (ABNT2015). Figure 1 shows the architecture of different service environments.

Fig. 1. Architectural images of different service environments

We observed that regulations fit the concept of ID, seeking access for all people. However, the physical spaces of the institution and the oncology service do not follow the parameters required to include all users, regardless of their needs. There was no signage that met the visual, sound, and tactile needs. That is, a user with reduced vision, or deaf, could face difficulties or be excluded from care.

When we asked about aspects of physical accessibility both for professionals and users, in the interviews, the answers always referred to spaces structured to receive wheelchair users. However, we know that accessibility is not just for wheelchair users.

Most of the inclusive environments have not been user-friendly in actual practice. The findings suggest that while policies are an essential precondition, implementation and management must not be ignored because they ensure the effectiveness of ID. One reason that justifies this finding is the fact that the premises that house the oncology service pre-dated the establishment of the currently existing LSP, requiring adjustments. Furthermore, it is essential to understand that an environment must be ready to receive users whatever their needs.

As regards **information and communication aspects**, data refer to any question, attitude, or behavior that promotes or hinders the expression or comprehension of messages and information. Laws 10.098 and 13.146 recommend the adoption of comprehensive forms of interaction, which include Brazilian Sign Language (Libras); the visualisation of texts in Braille; signaling or tactile communication systems, enlarged characters, etc. (Brazil 2000; Brazil 2015).

In practice, these aspects were rarely observed. We identified that communication and access to information are mostly carried out orally or in writing, disregarding the recommendations of LSP. Adopting clear communication and easy-to-understand terms is a good starting point for patient safety and effective communication. When the information is easy to understand and uses simple language, it improves the receiver's knowledge and satisfaction, helping to clarify doubts and develop reflective skills, to provide autonomy and adherence.

About **technological aspects**, data are those that promote or hinder people with vulnerability from accessing technologies, such as existing treatments, qualified teams, assistive technologies, and others. Laws 10.098 and 13.146 recommend the implementation of assistive technologies so that people with limitations have their access or use of services and spaces facilitated, with technologies adapted to their needs (Brazil 2000; Brazil 2015). In the oncology service observed, however, assistive technologies adapted to the needs of all people were not perceived nor was the application of any technology observed. This may be because the service analysed is from the SUS, despite its having a large amount of financial resources, sometimes it is not enough to cover the thousands of people who need to use it.

However, within the technological scope, the service provides access to the different existing treatments, as required by Law 14.238 (Brazil 2021). Despite this, it was found that patients have access to treatments covered by the SUS that are considered traditional. On the other hand, patients do not have a proper understanding of the rights guaranteed by legislation since we found that the decision-making about which treatment to adopt is largely defined by doctors.

The last category of analysis refers to **attitudinal aspects**, which means attitudes or behaviors that promote or harm the social participation of vulnerable people on equal terms with others. To reinforce human practice in the provision of health services, in 2003, Brazil established the National Humanization Policy (NHP), aimed at full access, universal and equitable spaces, in the daily life of health service users (Brazil 2004). Furthermore, Laws 13.146 and 14.238 ensure the promotion, under equal conditions, of the exercise of fundamental rights and freedoms by people with disabilities, with a view to their social inclusion and citizenship rights.

In the oncology service, humanised attitudes were reported mainly in situations of patients' interaction with the oncology service, i.e., medical, nursing, and administrative professionals, which was defined as good by patients. The humanised reception generates a relationship of trust and respect, increasing the feeling of well-being and a positive experience, corroborating Hashemi et al. (2022). However, situations that generated discomfort in the patients were observed, throughout the service. Following the guidelines of the Statute of the Person with Cancer, humanisation must be encouraged, and care for cancer patients must be aligned with the guarantee of physical, psychological, emotional, and social well-being (Brazil 2021).

4 Discussion and Conclusions

We identified that LSP can foster inclusive public policies and facilitate the inclusion of patients in health services, including at local levels. We noted the importance of establishing systematic top-down public policies related to ID. However, having comprehensive public policies is not enough to guarantee the effective implementation of the requirements. Implementation is an essential step in planning accessible and inclusive services, putting defined policies into practice. However, one of the difficulties identified in the service has been the non-implementation of inclusive services required by legislation to satisfy real patients. Finally, management is a fundamental resource that ensures the effectiveness of accessible and inclusive services, encompassing maintenance, evaluation, supervision, and participation.

Considering the context presented, we verified the establishment of Brazilian laws and norms that apply the principles of ID, aiming to promote accessibility and inclusion. The incorporation of ID principles has taken place in an evolutionary process over time, according to Fig. 2. We noted that the regulations fit in the concept of ID, seeking access for all people. The findings also failed to identify any practices in addition to what is provided for in the LSP.

From this case study, we can confirm some good practices. We emphasise the understanding that public policies related to ID are an essential base, to enhance social sustainability. In this sense, we recommend that the design of these laws and regulations includes participatory processes for different actors, aiming to obtain different perspectives on the same issues. Besides that, implementation and management are vital as they ensure the effectiveness of ID. So, long-term and ongoing actions at different levels of policy conception, implementation, and management must be taken to provide a more comprehensive understanding of ID in healthcare services. To inspire better practices of inclusion and accessibility in developing countries with public health systems, the main lesson learned is that health services must be in line with the country's public policies. Furthermore, there must be a dialogue between service providers and managers. Formulators are key players in guaranteeing rights, but they need to understand the needs of service practices and their users. In this sense, ID has much to contribute through a creative design process to promote the role of design in healthcare.

We understand that this study has limitations. First, it is focused on the laws and patient experiences, and future studies may expand the scope by bringing different professional views. Second, this case study described a particular service, so results are not generalisable. For future studies, participatory design processes involving all service stakeholders may help generate solutions to make the service more inclusive. In addition, quantitative studies are suggested, as well as comparative studies, to make results more generalisable.

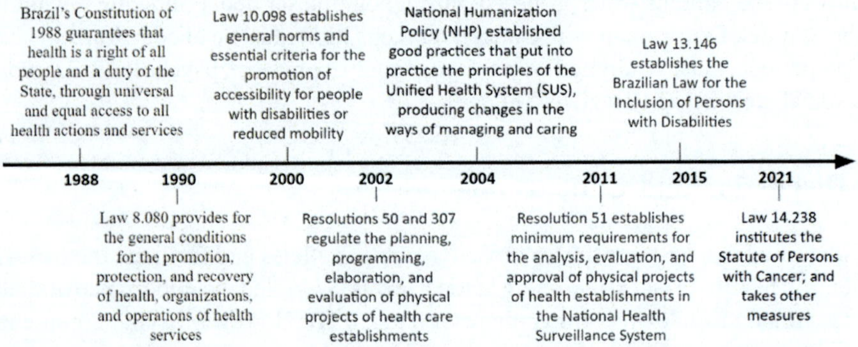

Fig. 2. Evolution of the relationship between public policies and inclusive design

References

ABNT: Norma Brasileira 9050. Associação Brasileira de Normas Técnicas (2015). http://acessibilidade.unb.br/images/PDF/NORMA_NBR-9050.pdf. Accessed 16 Jan 2023

Bohn, P.R., Konig, E., Souza, C.S.: Humanization in oncological health services: a Brazilian case study. In: 11th Inclusive Design Conference, London, UK, October 2022. CAAE 55467222.5.0000.5345

Braun, V., Clarke, V.: What can "thematic analysis" offer health and wellbeing researchers? Int. J. Qual. Stud. Health Well Being **9**, 26152 (2014)

Brazil: Constituição da República Federativa do Brasil de 1988 (1988). www.planalto.gov.br/ccivil_03/constituicao/constituicao.htm. Accessed 16 Jan 2023

Brazil: Lei nº 8.080, de 19 de setembro de 1990 (1990). www.planalto.gov.br/ccivil_03/leis/l8080.htm. Accessed 16 Jan 2023

Brazil: Lei nº 10.098, de 19 de Dezembro de 2000 (2000). www.planalto.gov.br/ccivil_03/LEIS/L10098.htm. Accessed 16 Jan 2023

Brazil: Resolução-RDC nº 50, de 21 de Fevereiro de 2002a (2002a). https://bvsms.saude.gov.br/bvs/saudelegis/anvisa/2002a/rdc0050_21_02_2002a.html. Accessed 16 Jan 2023

Brazil: Resolução-RDC nº 307, de 14 de Novembro de 2002b (2020b). https://bvsms.saude.gov.br/bvs/saudelegis/anvisa/2002b/rdc0307_14_11_2002b.html. Accessed 16 Jan 2023

Brazil: HumanizaSUS: Política nacional de humanização, p. 51. MS, Brasília (2004). https://bvsms.saude.gov.br/bvs/publicacoes/humanizaSUS_politica_nacional_humanizacao.pdf. Accessed 16 Jan 2023

Brazil: Resolução da Diretoria Colegiada - RDC nº 51, de 6 de Outubro de 2011 (2011). https://bvsms.saude.gov.br/bvs/saudelegis/anvisa/2011/rdc0051_06_10_2011.html. Accessed 16 Jan 2023

Brazil: Lei nº 13.146, de 6 de Julho de 2015 (2015). www.planalto.gov.br/ccivil_03/ato2015-2018/2015/lei/l13146.htm. Accessed 16 Jan 2023

Brazil: Lei nº 14.238, de 19 de Novembro de 2021 (2021). www.planalto.gov.br/ccivil_03/_ato2019-2022/2021/lei/L14238.htm. Accessed 16 Jan 2023

Carayon, P., Hundt, A.S., Hoonakker, P.: Technology barriers and strategies in coordinating care for chronically ill patients. Appl. Ergon. **78**, 240–247 (2019)

Costa, C.A., Lunardi Filho, W.D., Soares, N.V.: Assistência humanizada ao cliente oncológico: reflexões junto à equipe. Rev. Bras. Enferm. **56**(3), 310–314 (2003)

Fisk, R.P., Dean, A.M., Alkire (née Nasr), L., Joubert, A., Previte, J., et al.: Design for service inclusion: creating inclusive service systems by 2050. J. Serv. Manage. 2–5 (2018)

Garnelo, L., Parente, R.C.P., Puchiarelli, M.L.R., Correia, P.C., Torres, M.V., et al.: Barriers to access and organization of primary health care services for rural riverside populations in the Amazon. Int. J. Equity Health **19**, 1–14 (2020)

GDIH: Inclusive Design Strategy 2022. Global Disability Innovation Hub (2022). https://cdn.disabilityinnovation.com/uploads/documents/WUF-ID-Strategy-Draft-3-compressed.pdf?v=1656183741. Accessed 16 Jan 2023

Geraldo, J.P.B., Andrade, S.M.O.: Pessoas com deficiência e as barreiras aos serviços de saúde: uma metassíntese. Res. Soc. Dev. **11**(6), 1–16 (2020)

Hashemi, G., Wickenden, M., Bright, T., Kuper, H.: Barriers to accessing primary healthcare services for people with disabilities in low and middle-income countries, a meta-synthesis of qualitative studies. Disabil. Rehabil. **44**(8), 1207–1220 (2020)

Liu, Y., Dickerson, T., Early, F., Fuld, J., Jiang, C., et al.: Understanding the influences of COPD patient's capability on the uptake of pulmonary rehabilitation in the UK through an inclusive design approach. Int. J. Chron. Obstruct. Pulmon. Dis. **16**, 1717–1740 (2021)

Naughton, M.J., Weaver, K.E.: Physical and mental health among cancer survivors: considerations for long-term care and quality of life. N C Med. J. **75**, 283–286 (2014)

Persad, U., Langdon, P., Clarkson, J.: Characterising user capabilities to support inclusive design evaluation. Univ. Access Inf. Soc. **6**, 119–135 (2007)

UN: Many continue to face social exclusion and discrimination despite global social progress in recent years, warns new report. United Nations (2016). www.un.org/development/desa/en/news/social/rwss-launch-2016.html. Accessed 16 Jan 2023

How to Design for Everyday Designers? Learning by Redesigning a Child Oncology Ward

P. Tutenel[1](✉), A. Balmaseda Dominguez[2,3], and A. Heylighen[1]

[1] Department of Architecture, Research[x]Design, KU Leuven, Leuven, Belgium
{piet.tutenel,ann.heylighen}@kuleuven.be
[2] Faculty of Architecture and Urban Planning, University of Stuttgart, Stadtplanung und Entwerfen (SuE), Stuttgart, Germany
[3] Department of Architecture, Università degli Studi Roma Tre, Rome, Italy
alba.balmaseda@uniroma3.it

Abstract. Research that aims to support human-centred architectural design increasingly recognises people as everyday designers. What remains unclear, however, is how architects can design in ways that support everyday design. We addressed this question in a one-week course in architectural engineering. The course started from a real-world design brief – the redesign of a child oncology ward – and combined design exercises with lectures, discussions with various experts, observational exercises in a school, and watching a TV-documentary together to gain insight into the design of care environments, and children's everyday practices. As data we use individual and group student work and as themes for analysis we select lessons learned and design interventions suggested by the students in relation to the question we started from. In terms of implications we emphasise attention to people's involvement in practices as strategy in designing for everyday designers, integrating possibilities for user-feedback as part of design proposals in architectural education, and TV-documentaries as possible sources of information in designing for sensitive and hard-to-enter contexts like hospital wards.

1 Introduction

A fundamental understanding in human-centred design is that use cannot be predicted, and that people's interactions with the human-made world may differ from what was intended during the design process (Van der Bijl-Brouwer and Dorst 2017). Design research that aims to address the distance between design and use predominantly has two starting points: either 1) how design impacts on people; or 2) what people do with and in it (cf. Pink et al. 2020).

Studies in the field of architectural design focussing on the second starting point examine interactions between the built environment and its users and show how people consciously or accidentally redesign it through (everyday) practices (e.g. Sharif 2020). People are considered 'everyday designers' (Wakkary and Maestri 2008). In relation

J. Goodman-Deane et al. (Eds.): CWUAAT 2023, *Design for Sustainable Inclusion*, pp. 108–117, 2023.
https://doi.org/10.1007/978-3-031-28528-8_12

to architectural design practice, very few studies (if any) address the question of how architects can design in ways that support people as everyday designers.

This gap in research can be linked to a more general issue in architectural design, namely that design-use complexities – i.e., issues of agency and control – are rarely made explicit in design processes (Stam et al. 2020). Architects through design influence social life, but these outcomes can never be designed directly – they are deeply contingent (Redström 2008). Gaining insight into how design-use complexities figure during design processes and how architects deal with them is thus important. This contingency and its complexities related to design processes in architecture is particularly foregrounded in relation to children's interactions with the human made world (Jelić et al. 2020). More than most adults, children seemingly care less about designed intentions, and spaces/objects offer them possibilities for action that cannot be anticipated – the world of children appears to be less or undetermined (cf. Noens and Ramaekers 2014). In contemporary discourse about spaces for children, however, there seems to be a split between everyday life and care environments (Tutenel et al. 2022). In design (research), children in care environments tend to be considered as passive, acted on by these environments (more control, less agency), while children outside of care contexts are considered as active, able to act upon the environment (more agency, less control) (cf. Adams et al. 2017).

In view of these observations, we divide the question of how to design for everyday designers into three research questions: (RQ1) if we recognise that the design process continues after a design leaves the drawing board or when a building has been constructed, what does this mean for architects when designing?; (RQ2) how can architects design in ways that support children as everyday designers?; (RQ3) how does the idea of children as everyday designers question or reinforce current discourses and practices surrounding children in architectural design?

The study reported on here addresses these questions in the context of a one-week course in architectural engineering, which set off to rethink indeterminacy in design through exploring children's everyday practices.

The course was organised as part of a broader research project on how experiences of children affected by cancer may inform the design of cancer care environments (Tutenel 2021). With the redesign of a child oncology ward as an overall approach, the course combined concepts and lessons learned from the research project with design experiments in a series of seminars and hands-on design (and) research sessions.

2 Methodology

2.1 Course Set-Up

The course was organised in March 2022 in the framework of an international academic network – the Athens Network. The eight attending students – Benedetta Ballabio (IT), Daniele Florenzano (IT), Anna Nektaria Georgiou (GR), Theoni Gkesiou (GR), Nour Naim (LEBN), Jan Pospíšilík (CZ), Stavroula Tziourtzia (GR), and Vasco Pinto (PT) – were in their master's programme in Engineering sciences: Architecture. The authors of this chapter were the course organisers; an anthropologist, an architect and design tutor, and a design researcher all working in the field of inclusive architectural design.

As our overall approach to the research questions we worked by way of case study around the redesign of a child oncology ward in Belgium (Fig. 1). The design brief was explored through materials (architectural drawings and pictures taken by research participants) gathered in earlier research by a master student (De Boeck 2019). Now that the building has been inhabited for a few years, it turns out to deter some people from participation in practices, e.g. informal gathering.

The case was approached from multiple angles: informing, researching, exchanging, designing.

Students were **informed** through lectures and seminars. The third author linked a focus on children's practices to inclusive design. Starting from the understanding that 'disabled is not something one is, but something one becomes' (Moser 2005) she pointed out how human-made environments can either hinder or support children in their practices and that designers can create or resolve problems by creating disabling or enabling environments. The first author lectured about his research on children's everyday practices in an oncology ward (Tutenel 2021). He explained how conceptualising of children as everyday designers helped him to see these children not (only) as vulnerable but (also) as active constituents of the world (cf. Leeder-Mackley et al. 2015). Finally, the second author discussed with the students her approach to designing the *Padiglione Infanzia* by BarrioBalmaseda (Ariadna Barrio, Alba Balmaseda) and Esaú Acosta; an inclusive play space for children in Milan (IT) that represents the idea of universal design, which favors autonomy, active use, and comfort.

Research consisted of observational exercises. To become more attentive to children's everyday practices and their interactions with the environments they are in, students worked with school children for half a day through participant observation. Because we were not allowed to visit the hospital due to Covid measures, as a second observation exercise we watched a documentary together. Designers and researchers have already been experimenting with documentary filmmaking to inform designers. These design documentaries *'aim to get access to the elusive, inconsistent fabric of everyday life and understand it on its own terms'* (Raijmakers et al. 2006, p. 237). We decided to use, instead, a TV-documentary '*Pilotenmasker*' (de Jong 2017) about the daily life of children in an oncology ward, which was not specifically produced with the aim of informing design.

Next to observing children's everyday practices, **exchanging** with others was important. To get to know each other as (future) professional designers students introduced themselves by presenting one of their design projects. Another form of exchange was through online interviewing, consulting and discussing with various experts – an interior architect, a paediatric oncologist, a head nurse, a design researcher, and two young cancer survivors.

The students worked as a group on re**designing** the child oncology ward. A first exercise we call 'speed designing' (cf. Zimmerman and Forlizzi 2017) – a term borrowed from romantic speed dating. This exercise was the starting point of the redesign. Throughout the week students were given time to work on the redesign. During design time the first two authors were present to reflect with the students on their design considerations.

2.2 Material and Methods

We use as data for analysis in this chapter a video-recorded presentation in which the students jointly discussed online the course's trajectory and the design outcome with different experts, and the individual written assignments in which students reflect on what they have learned in relation to the aims of the course. This assignment (1000 words limit) was given as a take home exam and students had two weeks' time to hand it in.

Initially we had not intended to analyse the course, but the lessons learned from it in relation to the question(s) we started from motivated us to do so. Because of this we did not collect materials in a systematic way during the course's trajectory. However, to establish rigour in our analysis and presentation thereof in this chapter we limit ourselves to the above three types of materials – a recorded presentation, student design outcomes, and written student work.

As themes for the analysis we selected lessons learned and design interventions foregrounded by the students in relation to the question we started from. These were analysed in accordance with aspects of QUAGOL (Dierckx de Casterlé et al. 2012), which is a method guiding qualitative data analysis originally based on a grounded theory approach. Characteristic of this method is its extensive pen and paper phase planned to develop a holistic understanding of the data gathered. For our analysis, the first and second author independently developed initial concept schemes based on re-viewing the presentation and re-reading the student papers. These schemes were iteratively refined. Each step along the way was discussed among the three authors until consensus was reached.

As regards to research ethics, we distinguish between research ethics applied to running the course, and research ethics related to studying it. Concerning the former, De Boeck (2019) obtained informed consent from the architecture firm, the hospital research ethics committee (ref. BC-4807), and the children who participated in her Master thesis research for the materials used to explore the design brief. For the school visit, we asked permission via the school principal, who informed the children, parents and teachers of our observational exercise. Compliant with the General Data Protection Regulation we agreed with the school to not take pictures, only sketching and note taking were allowed. Ethical issues also come into play when using a TV-documentary as research material, e.g. to use it for a different purpose than was intended by the makers. In our course, we used '*Pilotenmasker*' (de Jong 2017) to learn about everyday life in a child oncology ward, which aligns with the purpose of the makers. Related to the analysis for this chapter we asked students' consent to use their assignment as data only after the evaluation and grading took place; and sent them a draft of the chapter before final submission. Consent for recording the presentation to and discussion with the experts was asked before the discussion started.

3 Findings

In answering the question of how to design for everyday designers, the students used the metaphor of a sponge to describe their design outcomes: like a sponge's infinite patterns and its ability to take many shapes, the design interventions aim at integrating open-endedness into the oncology ward to support children as everyday designers. First, we briefly discuss the students' design proposal, then we unravel their approach to better understand what can be learned in relation to our research questions (student quotes are in *Italics*).

3.1 The 'Sponge-Therapy' Design

The students agreed that **small things can be important**. Their redesign is made up of a series of micro-interventions (Fig. 1), which aim to create enabling moments for users and are easily adaptable to the existing environment (Fig. 2).

Important aspects were **communication and familiarity** in and through design. '*A simple, friendly, and to a certain extent familiar environment seems to be the way to go.*' A concern of the students was **to improve everyday life** in hospital environments, being '*able to help them [children] find those happy moments during the challenge they are facing with the disease*'. Finally, the **replicability** of the proposal in other medical care environments seemed relevant to them.

In order to obtain **feedback** on the design proposal, it was complemented with a survey for the users of the child oncology ward and an Instagram account.

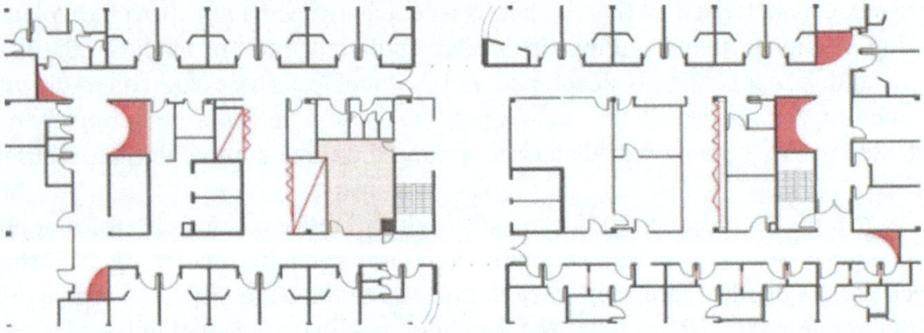

Fig. 1. Diagrammatic floor plan of the child oncology ward. The outer walls are omitted to respect non-disclosure. The students' micro-interventions are marked in pink. In response to the misuse of the informal meeting space (marked in light pink) the students proposed to specify their use through design while creating openness at other spaces in the ward. © Athens 2022 students.

Fig. 2. From left to right: '(Un)folding boxes' consist of storage spaces of different heights allowing storage of medical equipment on the upper parts, while children can benefit from lower cabinets for their toys, books or drawings. The exterior works as a skin where children can communicate with each other or make drawings; 'Capsules of Intimacy' are small spaces that can be used as private nooks; 'Tactility portals' make it possible for the children in isolation to interact with the children in the adjacent rooms; 'Mix and match' furnishes the parents' room/playroom with modular equipment. © Athens 2022 students.

3.2 Unravelling the Students' Approach

3.2.1 Becoming Aware of the Relevance of Seeing Children Differently

Approaching the case from multiple angles changed the students' ways of seeing children and questioned some of their conceptions of normality, inclusion, and familiarity. **Challenging normality** was triggered by recognising diversity among and different perspectives of children. In this reading, children are re-thought as having their own ways of participating in everyday practices that cannot be standardised or generalised. This showed in how the students 'came to see' children: *'They [the children] did not seem to worry too much about the obstacles, they just tried to solve the issues straight away in a simple way.'* And it showed how the students seemed to become more aware that children's perspectives are often narrowly defined (or forgotten) by adults: *'Many times adults focus on the playful side of children forgetting their very own personal perspective'*.

Another concept foregrounded by the students was **unfamiliarity**, which meant recognising that professional designers often know little about the spaces they design, about who the client is or the users will be. *'As designers, our role is to envision and deliver projects that affect and are used by "Others", but "Others" can mean a lot of things'*.

3.2.2 Understanding Children as Everyday Designers

Observing children in different contexts allowed the students to identify or recognise particular practices. At different occasions students referred to children as active agents. The notion of everyday design seemed to amplify this idea. This led to conceptualising children as having a natural tendency towards exploration, imagination and creativity: *'Observing children interacting between one another and [with] their environment but also with [them]selves was a great way to experience their infinite creativity.'* From this perspective, children create possibilities of using spaces and objects, provide different perceptions of the world and reveal unexpected patterns *'of adaptability and problem-solving'*. Observing children's ways of interacting with the school or the hospital environment helped the students to recognise different manifestations of children's everyday design practices in how they make and shape spaces for themselves or reinvent settings. To this end, children appropriate certain spaces, even if the space is not intended for them. *'They [children] treated unconventional surfaces such as glass windows as their own canvas* or *the moving bed as their own Ferrari.'*

In their reflections on the TV-documentary, students recognised how practices express the emotional experience of care environments, and the benefits of communication and contact to mitigate the feeling of vulnerability, loneliness when isolated, or the importance of being accompanied by familiar objects. *'I noticed that [...] even when they use the medical equipment, children were more relaxed when they knew exactly what was happening, when they were aware of the procedure and informed by the doctors.'*

3.2.3 Designing an Oncology Ward for Children as Everyday Designers

Reflecting on the design process students emphasised the importance of **direct interaction** with various experts – new to them. The conversation with one of the cancer survivors helped students to understand better what isolation in a hospital environment entails, what matters to hospitalised children, and what makes them feel better. Another activity of relevance to the process was carefully (closely) **observing children's everyday practices in care and other environments**. Next to interacting with and observing children, students found designing and researching together in **a network of people** (students, lecturers, experts, and children involved) and a transdisciplinary atmosphere was important. This provided the '*opportunity to investigate design using so many different approaches*'. In a structured and highly complex environment such as a children's oncology ward, the students foregrounded the **role of** ***'constraints*** *as a booster for creative and design process*'. Finally, in the design process, **feedback** from actual users was considered important to open up a horizon for future improvement. Students recognised that designing is a '*a vice versa procedure*' in which designers and users can mutually benefit.

3.2.4 Implications for Design Practice

Beyond this specific redesign the different angles seemed to stimulate reflection in the students about designers' **responsibility** more generally. The students' understanding of responsibility involved rethinking existing definitions and proposing solutions for specific uses and users. '*This project made me more empathetic, responsible, and aware of the impact we have on people's (especially aspiring young humans') everyday lives.*'

Next to these reflections, students described the need for a change in **attitude** towards the role of professional designers: '*The role of [an] architect will thus not be to strictly categorise the different areas of the hospital spaces as we have been used to, but rather to create opportunities for the interaction of different forms that can change over time. It will then be up to the children themselves to what extent they would wish to engage*'. On the one hand, a recognition that architects can contribute to create opportunities for interaction, on the other hand, the understanding that reality is ever changing, accepting that in the end the decision is the user's and that professional designers cannot meet or control everyone's needs. Related to these issues concerning agency and control students emphasised collaboration: for example, one student wrote that '*professional and everyday designers design together*', thereby situating design as an ongoing activity and process.

4 Discussion and Concluding Remarks

What does the students' redesign of a child oncology ward and their reflections on the course teach us about the research questions we started from? We learned, first, that to recognise that the design process continues is to see architects and everyday designers as designing *together*; it requires a change in attitude related to agency and control; and shows the potential of everyday practices as a starting point in design (cf. RQ1); second, designing in a way that supports children as everyday designers starts with taking children seriously and designing spaces that allow indeterminacy and are not separate from the fabric of everyday life, be it in a hospital ward, a shop, or on a pavement (cf. RQ2). And third, understanding children as everyday designers helped the students to recognise (balance) children's vulnerable situation when being hospitalised while keeping an eye on how even in such precarious situations children shape spaces for themselves; to understand their responsibility as professional designers to support these children's agency (cf. RQ3).

How do these preliminary findings relate to previous research, and what are their implications? To start with, research that focusses on how architects deal with design-use complexities in practice shows that strategies related to developing *specificity* include gaining insight into perspectives of (possible) users, discussing scenarios with clients, etc. (Stam et al. 2020) For the students who participated in our course, a strategy to cope with architecture's contingency precisely was to pay close attention to how children are involved in everyday practices in order to create *openness* in their design proposal. And, in relation to understanding children as everyday designers the students designed interventions that may support children as such not outside or at the margins (e.g. in the playroom) but right in the oncology ward, as part of ongoing everyday practices.

Second, although not common in architectural education, the participating students included possibilities for feedback from people with relevant expertise as an integral part of their design proposal. For them it was a logical next step in dealing with design's indeterminacy, complementing their proposal with a survey related to (possible) everyday uses and an Instagram account. The survey was the students' way of acknowledging that architects do not have (full) control over design outcomes and need feedback from a broad range of users to steer the proposal further in relevant directions; and that, in the end, it will be up to the children/users to what extent they wish to engage with the outcome. As such, integrating possibilities for feedback, like the students did, may be an important strategy for student-architects to deal with design-use complexities. Moreover, it may shed a different light on, for example, post-occupancy evaluations, understanding these as part of design processes and based on everyday uses (cf. Simonsen et al. 2022).

Finally, research suggests that professional designers are attracted to information presented with maximum use of graphics and limited text (Annemans et al. 2014). They prefer raw data in a format that is condensed to be design-relevant, allowing them both to quickly pick up overarching themes and to discover depth relevant to a specific project (McGinley and Dong 2009). In our course we worked with an existing TV-documentary. Watching it together using everyday design as lens helped students to focus on interactions between children and the material care environment. Our findings suggest possibilities of using TV-documentaries as one possible source of information in sensitive and hard-to-enter contexts like hospital wards.

An important limitation of the study is that because we did not originally intend to study and write about the course, the materials we could use for the analysis were limited – recorded and written work. Related to this, the reflections of the students were part of their evaluation, which means that the content of the reflection papers could be influenced also by e.g. wanting to achieve good marks. Thirdly, students suggest the need for a change in attitude towards the role of professional designers. Studying this would require an analysis of their (design) work done after the course is finished, which we did not do. Finally, since the students' design proposal was not realised and did not lead to interventions in the oncology ward, we cannot say much about whether and how it will support children as everyday designers. The discussion with the experts, however, brought to the fore points of attention in the students' design proposal related to practicalities and materialisation: not taking into account hygiene standards in an oncology ward, materials that can or cannot be used, extra maintenance work, etc.

We see three key areas for future research. The first follows from the circumstances that made the course: this was the first time we organised a course on children's everyday (design) practices that was both research-based and design-oriented, there were not many examples available that have followed this format which we could take as a reference, the context of the international academic network challenged us to introduce students to a broad and unfamiliar topic in just five days, and the course had to adapt to the constraints of the COVID-19 pandemic, which is why meetings with experts took place online, and the idea of using a TV-documentary arose. In light of these limitations a first area for future research would be a second iteration of the course, also because this would allow a more systematic gathering of materials. A second area is to further explore TV-documentaries as a way to inform design for sensitive contexts and develop an ethical framework that allows this. And finally, it would be interesting to extend the experience from students to professional designers, working on real-world design projects.

Acknowledgements. We thank the students, Pascale Conard and the Athens network, and Begga Willems and the *De Klare Bron*. Thanks also to Wout Gastmans, Geneviève Laureys, Annelies Mannaerts, Pieter Vandekerckhove, and Elfi Van der Borght for sharing their expertise. We received support from *Kom op Tegen Kanker*.

References

Adams, P.C.: Place and extended agency. In: Enfield, N.J., Kockelman, P. (eds.) Distributed Agency, pp. 213–220. Oxford University Press, Oxford (2017)

Annemans, M., Van Audenhove, C., Vermolen, H., Heylighen, A.: How to introduce experiential user data. In: Proceedings of DRS 2014, Umeå, Sweden, pp. 1626–1637, 16–19 June 2014

De Boeck, S.: The common rooms of a pediatric oncology ward from a child's perspective [Unpublished Master Dissertation]. KU Leuven, Belgium (2019)

de Jong, S.: Pilotenmasker [video 95 min. NL]. EO Ikondocs. www.2doc.nl/documentaires/2017/11/pilotenmasker.html. Accessed 5 Dec 2022

Dierckx de Casterlé, B., Gastmans, C., Bryon, E., Denier, Y.: QUAGOL. Int. J. Nurs. Stud. **49**(3), 360–371 (2012)

Jelic, A., Martin, M., Holst Laursen, L., Tvedebrink Doktor Olsen, T., Fich Brorson, L., Oehlwein, L.I.: Children, play, and the built environment (Report 3). CoC Playful Minds. Billund, Denmark (2020)

Leeder-Mackley, K., Pink, S., Moroşanu, R.: Knowing the world through your body. In: Seymour, J., Hackett, A., Procter, L. (eds.) Children's Spatialities: Embodiment, Emotion and Agency – Studies in Childhood and Youth, pp. 21–38. Palgrave, London, UK (2015)

McGinley, C., Dong, H.: Accessing user information for use in design. In: Stephanidis, C. (ed.) UAHCI 2009. LNCS, vol. 5614, pp. 116–125. Springer, Heidelberg (2009). https://doi.org/10.1007/978-3-642-02707-9_13

Moser, I.: On becoming disabled and articulating alternatives. Cult. Stud. **19**(6), 667–700 (2005)

Noens, P., Ramaekers, S.: The family as a "gathering". Int. J. Child Youth Family Stud. **5**(4.2), 722–740 (2014)

Pink, S., Duque, M., Sumartojo, S., Vaughan, L.: Making spaces for staff breaks. HERD: Health Environ. Res. Des. J. **13**(2), 243–255 (2020)

Raijmakers, B., Gaver, W.W., Bishay, J.: Design documentaries. In: Proceedings of the 6th Conference on Designing Interactive Systems, University Park, PA, USA, pp. 229–238, 26–28 June 2006

Redström, J.: RE: definitions of use. Des. Stud. **29**(4), 410–423 (2008)

Sharif, A.A.: Users as co-designers. Front. Archit. Res. **9**(1), 106–118 (2020)

Simonsen, T., Sturge, J., Duff, C.: Healing architecture in healthcare. HERD: Health Environ. Res. Des. J. **9**, 19375867211072513 (2022)

Stam, L., Verbeek, P.P., Heylighen, A.: Between specificity and openness. Des. Stud. **66**, 54–81 (2020)

Tutenel, P.: Room for vulnerability [Unpublished doctoral dissertation]. KU Leuven, Belgium (2021)

Tutenel, P., Ramaekers, S., Heylighen, A.: The pavement and the hospital bed. J. Inter. Des. **47**(4), 3–10 (2022)

Van der Bijl-Brouwer, M., Dorst, K.: Advancing the strategic impact of human-centred design. Des. Stud. **53**, 1–23 (2017)

Wakkary, R., Maestri, L.: Aspects of everyday design. Int. J. HCI **24**(5), 478–491 (2008)

Zimmerman, J., Forlizzi, J.: Speed dating. She Ji **3**(1), 30–50 (2017)

Study of a Smartphone App as a Bridge Assistive Technology for a COVID-19 Home Test: 19 Essential Guidelines

S. K. Oswal[1](✉) and H. K. Oswal[2]

[1] Center for Research and Education on Accessible Technology and Experiences, School of Interdisciplinary Arts and Sciences, University of Washington, Seattle/Tacoma, WA, USA
oswal@uw.edu

[2] Science and Math Institute, Tacoma, WA, USA

Abstract. This paper presents the results of an expert usability evaluation study focusing on the accessibility, medical exigency, and user independence of a COVID-19 Home Testing Kit that employed a smartphone app for communication. The paper also proposes a set of guidelines for designing medical testing kits employing smartphones as a bridge technology. Researchers have argued for co-design approaches for usable technology for disabled consumers. Acknowledging disabled users as experts of their contexts of use and using their valuable lived experiences with day-to-day technologies are essential steps in inclusive design. Engaging design researchers' experiences of working with disabled users, we revisit these challenges in the context of this kit to evaluate the effectiveness both for accessibility and usability. Since our test discovered several design issues with the interaction design of the kit and the assistive technology embedded in the testing process, we argue for more comprehensive models for designing medical testing kits that would integrate both assistive technologies and other elements seamlessly and accessibly. The designer/engineer's ability to view technology design as one element in a larger usability ecosystem, including nontechnological factors, is central to an independently performable home test. Other important factors are: designer/engineer's familiarity and comfort with disability, prior experience of working in a disabled community, use of similar existing assistive technology designs for this user group, and an understanding of the complexity of the users' environment where technology is employed.

1 Introduction

The international standard ISO 9241-210 describes user experience as: "a person's perceptions and responses that result from the use or anticipated use of a product, system or sendee" (ISO 2019). This standard is an extension of ISO 9241-11 which deals with the concept of usability (ISO 2018). Whereas usability focuses on measuring the effectiveness and efficiency of a product, process, or system in accomplishing tasks, user experience (UX) is dynamic, context dependent, and among others includes user satisfaction in its widest terms, ranging from the response to related factors—how quickly the product has been shipped, how difficult the package was to open, and the user emotions aroused by their interactions with all aspects of product use (Law et al. 2009).

J. Goodman-Deane et al. (Eds.): CWUAAT 2023, *Design for Sustainable Inclusion*, pp. 118–126, 2023.
https://doi.org/10.1007/978-3-031-28528-8_13

Research on the application of these standards to the design of medical diagnostic tests in disability contexts is not yet common.

Six months after the US government made free COVID-19 home tests available to citizens, the U.S. president announced the availability of an accessible COVID-19 home test kit for the blind by Ellume on 23 June 2022. Using the case of this Ellume COVID-19 home test kit—sometimes also labeled as audio test for the blind on the manufacturer's website—we underscore the technology gaps in the kit design. We chose to test the Ellume home test for two reasons: 1) Ellume is made available by the US government free of cost to the blind and 2) the only other test with a claim of accessibility by the blind—Cue—sells for $450 and up and is beyond the reach of most blind users, a majority of whom live on disability income. After testing the usability and accessibility of the Ellume kit, the primary author of this paper reached out to the company for their Voluntary Product Accessibility Template (VPAT) for this product in July. The company did not have a VPAT available (Personal email 13 July 2022). We might note that Section 508 of 1998 of the revised U.S. Rehabilitation Act of 1973 requires suppliers to provide a VPAT for assuring product accessibility (Section 508.gov).

2 Relevant Literature Review

Researchers have highlighted the centrality of assistive technology to vulnerable populations, particularly to the disabled or older adults, during the COVID-19 pandemic. Isolation, fear of infection, and lockdowns took away most support systems from this class of users and when assistive technologies did not work, or were not available, these groups suffered more than the rest of the population (Jutai and Tuazon 2022). Disabled people in general had difficulty receiving care in times of health crisis during this pandemic (Guidry-Grimes et al. 2020; Brown et al. 2022).

A primary purpose of universal design (UD), inclusive design, and accessible design approaches is to make consumer products, interactive technologies, healthcare services, and environments available to a variety of users with and without disabilities (Milne et al. 2014; Persson et al. 2015). Universal design is based on Ronald Mace's barrier-free and accessible design principles that advocate for access to all environments (Mace et al. 1991; Connell et al. 1997; Vanderheiden 1998). ISO's guide 71 defines UD as "design focused on principles of extending the standard design to persons with some type of performance limitation to maximise the number of potential customers who can readily use a product, building or service [that] is usable by most users without any modification" (ISO/IEC 2014).

3 Methods

This study used a basic usability test with accessibility for blind users in mind (Dumas and Redish 1999). We conducted two expert tests with an observer taking notes. The primary author served as the expert tester and the second author played the role of an observer and note taker. The expert tester was familiar with both Windows based screen readers and iOS VoiceOver screen reader; had previous experience of testing physical equipment and digital interfaces; and could perform testing steps without using sight.

The usability and accessibility test was repeated twice to capture data in depth and to detect any overlooked details. We split our original notes into two parts, usability and accessibility test of the smartphone app and the actual COVID-19 test. We recorded our testing session for later reference during the analysis (Table 1).

Table 1. Questions guiding our study

Criteria for Testing the Kit
1. Is the website for ordering the COVID-19 Test accessible?
2. Is the COVID-19 Test package designed for a blind user's convenience?
3. Are the various components of the COVID-19 test easy to unpack and identify with or without accompanying instructions?
4. Is simple and straightforward language used for labeling and describing equipment, explaining the procedure, and giving step-by-step instructions?
5. Are the warnings and alerts given at the appropriate point in the test?

4 Description and Use of Ellume COVID-19 Home Test Equipment

Mobile application (App)-based COVID-19 at-home self diagnosing test kits have been introduced in the U.S. market widely although stand-alone analysers without a mobile platform are more common (Zai et al. 2022). Therefore, for the purpose of this study, we ordered 12 Ellume tests which arrived in a large box with a print-only address label. This test kit consists of an analyser with a Bluetooth reader (the testing unit), a collection tube which Ellume labels as a dropper, a vial of processing fluid, a nasal swab for taking samples, a quick start guide, a 9-fold, two-sided product information leaflet, and a QR code printed on the right panel of the kit box. The box does not include a biohazard bag for disposal of the kit after use. The manufacturer provides a downloadable app for iOS and Android smartphones that communicates results from the analyser through a Bluetooth connection.

4.1 What is Involved in Performing the Ellume COVID-19 Home Test

The steps for completing this home test can be split in three phases: 1) downloading the mobile app on the user's smartphone and entering user data, 2) performing the COVID-19 test using the analyser, and 3) reading the results from the mobile app.

5 Results

Overall, the testing kit and the app were easy to use and navigate visually. However, when we tested the kit solely depending on the spoken directions from the iOS app, other sounds, such as beeps and blips, and touch to conduct the test, the whole procedure was not only tedious but also patchy due to the accessibility gaps in the spoken directions. Textual or audio descriptions of buttons and procedural steps in this app lack a meaningful structure. In addition, several buttons in the app were mislabeled or not labeled at all for the iOS screen reader, VoiceOver. Some of the instructions were poorly worded, the video narration was slow, and the information in the animations had no alternative text. The kit provided no accessible description of the equipment, and a blind user will have to guess which part is what. The individually packaged parts had few or no tabs and would be difficult to open for anyone with arthritis.

During the two expert tests, we identified a number of accessibility barriers for the blind. In addition, we also found several usability problems even when the test was otherwise accessible. The video instruction manual provides little information of the location of the on off button on the analyser. Once the video has been started, it can not be paused. The video shows animated images, but the oral instructions do not describe the imagery; the oral instructions are minimal and require some imagination. When repeating the test, the user is forced to watch the video again because the app has no skip button. The video does have a pause button, but it is not labeled for the screen reader. The run test label reads only as "next button". The 15-min timer for processing the sample has no audio notifications. When the app displays the test results, the only automatic announcement is "IMG menu"—read by VoiceOver because this image is unlabeled. The blind user has to use touch to have the screen reader read the results instead of the app automatically announcing them (Table 2).

Table 2. Basic steps a blind user might take to perform the COVID-19 home test

Workflow for Completing the Ellume Home Test
• After opening the Ellume kit box, the blind user reviews print instructions with sighted help or searches for the same online • The user locates the unmarked QR code on the kit box and downloads the Ellume app on a smartphone • The user enters required and optional information in several steps in the partially accessible Ellume app • The user watches mandatory video with directions for performing test from the app • The user turns on the analyser and connects it to the Ellume app • The user transfers processing fluid from the sealed vial to the dropper • The user collects a nasal sample using the swab • The user screws the nasal swab onto the dropper • The user squeezes out 5 drops of processing fluid from the dropper into the analyser sample port • The user presses the Run Test button on the app and the analyser starts processing the test sample • After 15 min, the smartphone app displays the test results

6 Discussion

This accessible kit was issued during summer 2022 when the COVID-19 pandemic had already been pushed out of the public consciousness by most of the federal, state, and local government agencies in the United States having eliminated most safety requirements—masks, social distancing, and testing requirements. For blind individuals with other underlying conditions and at a higher risk for COVID-19 infection, getting a reliable home COVID-19 test is still a basic need. With most safety mechanisms gone, these individuals are at a greater risk in public and sheltering at home has been their best mechanism for self protection. Blind users had also stressed to the U.S. government the transportation barriers to accessing these tests from healthcare facilities (National Federation of the Blind 2022). No doubt, they were also at a greater risk of exposure in traveling to such sites. By the time this semi-accessible kit was made available, other people in U.S. had easy access to free home tests although this access also declined after 3 September 2022, when the Federal Government stopped funding COVID-19 home testing kits across the board.

It is hard to determine if the gaps in product accessibility are simply due to the shortage of experienced accessibility designers, or due to the emergency conditions under which the design work on these home testing kits had to be undertaken. In either case, it is difficult to rationalise the accessibility gaps in this audio COVID-19 Test Kit because it was announced by the U.S. Government specifically to meet the demand for an accessible home test kit for the blind.

Although disability activists have stressed the need for inclusive designs, in medical contexts a universal design for all users may not serve everyone's needs (Charlton 1998). To address specific disability contexts, researchers have advocated ability-based designs—designs that cater to the specific needs of users with specific abilities (Wobbrock et al. 2011). When we are thinking of making the affordances of our technology accessible, as well as usable for consumers with disabilities, the concept of accessibility has to be complicated for the sheer reason that the contexts-of-use for this population might have additional variables requiring added affordances. These essential affordances might vary from disability to disability. In considering these users' needs, the designers also need to pay attention to how these differing contexts-of-use are reshaping the nature of the desired activity. Further, we need to pay attention to how the revised activity affects the expected user experience.

The context-of-use in the case of the COVID-19 home test is not complex in the traditional sense of the term because the testing task has a well-defined purpose—checking whether or not the user is presently testing negative or positive, has a clear-cut procedure to perform the test, and involves a set number of variables for performing the test. However, from the disabled user's perspective, the task is more complex because they have to accommodate their body to the procedural needs of a test designed for visual users. Since the kit has quite a few images explaining different aspects of the test, sighted users can easily follow the procedure. On the other hand, a blind user might be altogether unfamiliar with the basic equipment required for testing and managing the relevant information to complete the test.

Because a sighted user also receives cues from the layout and visual design of the app, the step-by-step instructions often assume certain details. When the step-by-step instructions are interrupted by warnings and alerts, an aural user is repeatedly distracted from the procedural task. To address these issues and to enable the user to complete the medical test with a high degree of accuracy, the user instructions need to anticipate the inexperienced users' often asked questions and answer them adequately.

To understand how disabled people use medical devices and procedures, it might not be enough to employ generic usability evaluation methods. Seeking out disabled expert users of such contexts might be essential to gain insights into how these users perform complicated tasks successfully with the available sensory affordances and compare their techniques to the less experienced users' ways of approaching the task.

Researchers have also emphasised the importance of providing strategic information for designers to develop accessible technologies (Goodman et al. 2006). We further emphasise that codesigning and testing prototypes with actual users does assist designers to identify user concerns and pain points early on in the design stage and to come out with a product that matches users' needs and expectations (Hall 2001; Oswal 2014; Shinohara et al. 2016).

7 Conclusion

Overall, the usability and accessibility tests suggested that the kit did not meet the needs of blind users. Making a medical test technically accessible is not sufficient, it also has to possess the general usability affordances available to visual users. It is also not clear if any input was received from blind users in the early stages of design; otherwise, this product should not have had the obvious design issues we identified. Considering the barriers posed in completing this diagnostic test without sight, the claim that this COVID-19 home test kit provides independent access for blind users is untenable (White House COVID-19 Response Team 23 June 2022). Nevertheless, the availability of a free test to the blind with some accessibility has opened a door for further improvements.

The lessons learned from this study of a COVID-19 home test can also be exceptionally valuable in other fields where disabled users require autonomy in performing technical procedures. Our study also highlights the need for engaging disabled users in the design process from early on and throughout the product's development cycle. Further, it draws attention to the need for designing more effective interfaces between medical equipment and the digital apps employed for communicating user instructions and test results.

We present a set of guidelines in the table overleaf (Table 3) for designing medical kits of this nature that offers strategic information for designers and developers to conceptualise accessibility and usability for blind users. These guidelines answer a crucial question: "How do you design for 'best practices' for accessibility in medical diagnostic tests and related procedures?".

Table 3. A set of guidelines for designing an accessible home test for the blind

19 Guidelines for Designing a Home Medical Test Kit
1. Think of the whole ecosystem within which a blind user will handle the equipment and perform the test
2. Apply universal design (UD) principles across the board when designing test equipment
3. Apply Web Content Accessibility Guidelines for mobile apps to the digital interface of the test
4. Ensure that the website for ordering the kit is accessible, including its images
5. Pay attention to the size and shape of parts and tools at the design stage to meet the unique needs of blind users
6. Test the prototype with blind users to seek feedback at the concept-forming stage because retrofitting accessibility after the fact is costly and often less effective (Wentz et al. 2011; Horton and Sloan 2014; Oswal 2019)
7. Make the various components of the diagnostic test easy to unpack and identify with or without accompanying instructions
8. If using a mobile app for communication, integrate an audio signal system in the analyser to show test status
9. Provide a list of equipment parts and label each part in braille and large print
10. Include use scenarios for diverse user settings
11. Present step-by-step instructions with clues for confirming the accurate performance of each step
12. Make the sequence of steps for performing the test as predictable as possible and draw from blind users' existing experiences with technology
13. Test the instructions to assess the burden they place on the user (Suh et al. 2016)
14. Provide auditory or haptic landmarks when a particular task is completed successfully
15. Prepare users ahead of time with cautions and warnings to avoid mistakes
16. Place specific warnings and alerts at appropriate points in the test
17. When using digital interfaces, employ both large print and audio cues to support the needs of blind and visually impaired users
18. Create a troubleshooting section in the user manual that covers the salient pain points experienced by blind testers during user testing
19. Test FAQs with blind users to ensure that this information serves their needs

References

Brown, H.K., Saha, S., Chan, T.C., Cheung, A.M., Fralick, M., et al.: Outcomes in patients with and without disability admitted to hospital with COVID-19: a retrospective cohort study. Can. Med. Assoc. J. **194**(4), E112–E121 (2022)

Charlton, J.I.: Nothing About Us Without Us. University of California Press, Berkeley (1998)

Connell, B.R., Jones, M., Mace, R., Mueller, J., Mullick, A., et al.: The principles of universal design. North Carolina State University, The Center for Universal Design, Raleigh, NC, USA (1997)

Dumas, J.S., Redish, J.: A Practical Guide to Usability Testing. Intellect Books, Bristol (1999)
Goodman, J., Langdon, P.M., Clarkson, P.J.: Providing strategic user information for designers: methods and initial findings. In: Clarkson, P.J., Langdon, P., Robinson, P. (eds.) Designing Accessible Technology, pp. 41–52. Springer, London (2006). https://doi.org/10.1007/1-84628-365-5_5
Guidry-Grimes, L., Savin, K., Stramondo, J.A., Reynolds, J.M., Tsaplina, M., et al.: Disability rights as a necessary framework for crisis standards of care and the future of health care. Hastings Cent. Rep. **50**(3), 28–32 (2020)
Hall, R.R.: Prototyping for usability of new technology. Int. J. Hum Comput Stud. **55**(4), 485–501 (2001)
Horton, S., Sloan, D.: Accessibility in practice: A process-driven approach to accessibility. In: Langdon, P.M., Lazar, J., Heylighen, A., Dong, H. (eds.) Inclusive designing, pp. 105–115. Springer, Cham (2014). https://doi.org/10.1007/978-3-319-05095-9_10
ISO/IEC: ISO/IEC Guide 71:2014 Guide for addressing accessibility in standards. ISO, Switzerland (2014)
ISO: ISO IS 9241-11:2018 Ergonomics of human system interaction - Part 11: Guidance of usability. ISO, Switzerland (2018)
ISO: ISO IS 9241-210:2019 Ergonomics of human system interaction - Part 210: Human-centred design for interactive systems. ISO, Switzerland (2019)
Jutai, J.W., Tuazon, J.R.: The role of assistive technology in addressing social isolation, loneliness and health inequities among older adults during the COVID-19 pandemic. Disabil. Rehabil. Assist. Technol. **17**(3), 248–259 (2022)
Law, E., Roto, V., Hassenzahl, M., Vermeeren, A., Kort, J.: Understanding, scoping and defining user experience. In: Proceedings of the SIGCHI Conference on Human Factors in Computing Systems, Boston, MA, USA, 4–9 April 2009, pp. 719–728 (2009)
Mace, R.L., Hardie, G.J., Place, J.P.: Accessible environments: toward universal design. In: Wolfgang, F.E., Preiser, J.C., White, E.T. (eds.) Design intervention: Toward a more humane architecture, pp. 155–176. Van Nostrand Reinhold, New York (1991)
Milne, L.R., Bennett, C.L., Ladner, R.E.: The accessibility of mobile health sensors for blind users. In: International Technology and Persons with Disabilities Conference Scientific/Research Proceedings (CSUN 2014), San Diego, CA, USA, pp. 166–175 (2014)
National Federation of the Blind: Ellume COVID-19 home test evaluated for accessibility on 4 February 2022 (2022). https://acl.gov/sites/default/files/2022-07/COVID%20Test%20Accessibility%20Review%20-%20Ellume%20-%20NFB%20-%20508checked.pdf. Accessed 18 Jan 2023
Oswal, S.K.: Participatory design: barriers and possibilities. Commun. Des. Q. Rev. **2**(3), 14–19 (2014)
Oswal, S.K.: Breaking the exclusionary boundary between user experience and access: steps toward making UX inclusive of users with disabilities. In: Proceedings of the 37th ACM International Conference on the Design of Communication, Portland, OR, USA, 4–6 October 2019, pp. 1–8 (2019)
Persson, H., Ahman, H., Yngling, A.A., Gulliksen, J.: Universal design, inclusive design, accessible design, design for all: different concepts—one goal? On the concept of accessibility—historical, methodological and philosophical aspects. Univ. Access Inf. Soc. **14**(4), 505–526 (2015)
Section508.gov: Voluntary product accessibility template (VPAT®). Reviewed/updated in May 2022 (2022). www.section508.gov/sell/vpat/. Accessed 18 Jan 2023
Shinohara, K., Bennett, C.L., Wobbrock, J.O.: How designing for people with and without disabilities shapes student design thinking. In: Proceedings of the 18th International ACM SIGACCESS Conference on Computers and Accessibility, Reno, NV, USA, 23–26 October 2016, pp. 229–237 (2016)

Suh, H., Shahriaree, N., Hekler, E.B., Kientz, J.A.: Developing and validating the user burden scale: a tool for assessing user burden in computing systems. In: Proceedings of the 2016 CHI Conference on Human Factors in Computing Systems, San Jose, CA, USA, 7–12 May 2016, pp. 3988–3999 (2016)

Vanderheiden, G.C.: Universal design and assistive technology in communication and information technologies: alternatives or complements? Assist. Technol. **10**(1), 29–36 (1998)

Wentz, B., Jaeger, P.T., Lazar, J.: Retrofitting accessibility: the legal inequality of after-the-fact online access for persons with disabilities in the United States. First Monday **16**(11) (2011)

White House COVID-19 Response Team: Press briefing by White House COVID-19 response team and public health officials (2022). www.whitehouse.gov/briefing-room/press-briefings/2022/06/23/press-briefing-by-white-house-covid-19-response-team-and-public-health-officials-86/. Accessed 18 Jan 2023

Wobbrock, J.O., Kane, S.K., Gajos, K.Z., Harada, S., Froehlich, J.: Ability-based design: Concept, principles and examples. ACM Trans. Access. Comput. (TACCESS) **3**(3), 1–27 (2011)

Zai, A.H., Caffrey, M., O'Brien, C.S., Jiang, D., Kanyat, R., et al.: The Implementation and acceptability of a combined mobile application with a COVID-19 at-home test kit. Telemed. e-Health **28**(7), 1064–1069 (2022)

Assistive Technology

What to Consider When Exploring Multidimensional Stigmatisation and Assistive Technology Inclusive Design

E. Healy[(✉)]

Interaction Design Centre, Department of Computer Science and Information Systems, University of Limerick, Limerick, Ireland
eileen.healy@ul.ie

Abstract. This paper provides a narrative review of dyslexia literature to illuminate some broader topics to consider when designing assistive technology for adults with dyslexia. It introduces the fundamental underlying concerns of adults with dyslexia who face stigmatisation and discrimination, both real and perceived, at work. In doing so, it provides insight into how multidimensional stigmatisation is experienced and its consequences for assistive technology use and design. As a way to avoid existing pitfalls, a co-design approach is introduced to explore and define solutions to address multidimensional stigmatisation, including assistive technology use and design for research work. For designers and researchers, this means starting with the correct problem. Collaborating, exploring, and taking action helps construct a picture, interpret the real issues, and determine if progress is being made in a better direction. Ongoing and future work focuses on preliminary aspects to evaluate assistive technology-related stigma for doctoral students with dyslexia. The study intends to develop a standardised scale to measure perceived multidimensional stigmatisation among doctoral students with dyslexia.

1 Introduction

There is a good chance that Human-Computer Interaction (HCI) designers may have had a misguided their approach to assistive technology design in the workplace for adults with dyslexia. The fear of stigmatisation and its effects on assistive technology users has been overlooked due to unnoticed clues that should have been noticed. It is easy to assume that stigma is not a design issue and is more a societal concept (Shinohara and Wobbrock 2011). However, designers need to go beyond surface understanding to see all the different factors that impact everyday interactions. It is important to develop a non-reductionist perspective when working towards correct solutions to address unmet needs of adults with dyslexia, identify root causes, and understand users' priorities when using assistive technology (Shinohara and Wobbrock 2011). By thinking in systems, designers can gain more insight into the people they are designing for in their daily life (Meyer and Norman 2020).

J. Goodman-Deane et al. (Eds.): CWUAAT 2023, *Design for Sustainable Inclusion*, pp. 129–136, 2023.
https://doi.org/10.1007/978-3-031-28528-8_14

2 Background

2.1 Dyslexia Complexities

Globally, dyslexia is the most prevalent learning disability; in some instances, it presents difficulty with reading, decoding written text, and spelling (Snowling et al. 2020). The core challenges of poor reading, writing, and spelling persist into adulthood, yet debate continues over adult reading comprehension struggles (Moojen et al. 2020). As such, there is still no agreement about definitions, and our understanding has been noted as unclear but evolving; despite all efforts, dyslexia remains complex (Protopapas 2019). Comorbidities such as hyperactivity disorder (ADHD), dyspraxia, dyscalculia, dysgraphia, and emotional and social effects may also be present; (Moll et al. 2020); however, dyslexia is a standalone reading disorder.

The diverse language used to describe dyslexia, competing theories, expanding definitions, and a lack of clear diagnostics complicate our understanding (Snowling et al. 2020). Defining it remains controversial because the term is not universally accepted (Kirby et al. 2020). Questions have been raised about the credibility and unscientific understanding of dyslexia and the lack of scientific distinction between dyslexia and poor reading (Elliott 2020). Yet approaches based on pitting poor readers against students with dyslexia and the dyslexia myth are problematic (Ramus 2014). In sum, the repeated focus on an exact definition may lead to researchers ignoring that the root causes of dyslexia are multiple (Protopapas 2019).

In this context, it is useful that these opposing voices and shared ideas are understood as parts of a complex interactive system. Addressing these tensions through co-design supports inclusion that may drive some improvement (Meyer and Norman 2020).

2.2 Dyslexia in the Workplace

Most adults with dyslexia describe a fear of real and perceived stigmatisation and discrimination in the workplace, including medical doctors, police officers and staff, academics, and nurses (Shaw and Anderson 2017; Macdonald and Cosgrove 2019; Major and Tetley 2019). When non-disclosure is chosen, access to reasonable accommodations such as assistive technology is at risk, although there is uncertainty surrounding the effectiveness of these adaptations (Locke et al. 2017). However, disclosure is a personal choice with no legal requirements to disclose (Kinsella et al. 2017). Additional perceived barriers include stereotyping, a lack of understanding of dyslexia, questioning work competencies, threats to career progression and job security (Macdonald and Cosgrove 2019). Furthermore, individual identities and self-concepts can range from perceived superpowers to disability to non-identification (Doyle 2020).

Despite our growing understanding of dyslexia, there may still be much to learn about it in the workplace (Norstedt 2019; Hamilton 2020). What is evident is that most people's invisible disabilities remain invisible at work, including in academia (Syma 2019).

For designers, understanding the fundamental issues and the user context early on is essential, as is understanding that real problems often stem from multiple root causes. Given that most adults with dyslexia have expressed fears of stigma, it would be a mistake to ignore its possible impact on assistive technology use and design in academia. By ignoring the perceived stigma, designers may be designing not for people with dyslexia but instead for their own assumptions about what should work (Norman and Stappers 2015). Moreover, a challenge for designers is asking the right questions; a good question may illuminate underexplored issues that concern users.

2.3 Challenges for Human-Computer Interaction Designers

Adults with dyslexia should not be viewed as a homogenous group (Johansson et al. 2021). Individual differences and identities, language and cultural diversity, and wide-ranging emotional consequences are prominent (Livingston et al. 2018). Besides, lived experiences and emotional consequences are widely individual (Nalavany et al. 2018). The understudied emotional effects of dyslexia, together with the lack of knowledge of some HCI designers about emotional design, as well as teaching emotional design for dyslexia, are also problematic limitations (Crabb et al. 2019). Additionally, some professionals report a lack of understanding of accessibility guidelines for dyslexics (e.g. WCAG 2.0 and 2.1) (Pichiliani and Pizzolato 2019). Another consideration is that a one-size-fits-all approach to assistive technology design will not work for this user group; customisation is essential (Schröppel et al. 2021).

There has been criticism of tokenism and marginalisation of scholars with disabilities in assistive technology design. Some are calling for an end to the perception of them as lacking credible knowledge about their everyday circumstances (Ymous 2020). Yet, despite best efforts, a lack of recruitment persists for scholars with disabilities including adults with dyslexia in HCI studies (Motti 2019).

Based on these insights, it appears desirable that HCI, dyslexia, co-design, and stigma experts work with doctoral students with dyslexia to balance different perspectives, improve human agency, better unpack lived experience and collaboratively deconstruct assumptions, whilst focusing on users' real-world settings.

2.4 Universal Design, Stigma and Assistive Technology Design Gaps

The Universal design (UD) principle of equitable use specifies that designers should avoid segregating or stigmatising users (NDA 2020). A clear focus is placed on equity, autonomy, and digital inclusion. Despite the desire to provide universal access, design access for all approaches remains controversial in attempts to remove all obstacles in the real world for people with disabilities (Shakespeare 2013). He argues that achieving universal access is nearly impossible because it implies removing all barriers for all disabilities and degrees of the same disability.

Within this space, there is a need to consider the effects of stigma on an already stigmatised user group which is using stigma-sensitive interventions (Schröppel et al. 2021). For example, the evidence suggests that multidimensional stigma includes public stigma, self-stigma, cultural stigma, product-related stigma, and structural stigma (Vaes et al. 2016; Desmond et al. 2018; Macdonald and Cosgrove 2019; Jacobs et al. 2022). As it turns out, perceived stigma triggers assistive technology abandonment and may betray the invisible identity of the user (Schröppel et al. 2021). The paradox is that assistive technology users face stigma in both accepting and rejecting their technologies (O'Brolcháin and Gordijn 2018).

A tension between mainstream technology, accessibility and stigma issues for users with dyslexia was noted during the COVID-19 pandemic. For example, some online meeting tools do not offer text-to-speech functionality (Das et al. 2021). Against this backdrop, a lack of understanding of stigma-free design can lead to misleading assumptions and unintentional exclusion of users (Schröppel et al. 2021).

3 Methods

A realistic review (ongoing) approach was used to synthesise preliminary evidence and explain how interventions work or fail to work, for whom, how and in what circumstances (Wong et al. 2013). It explores the assumptions behind how and why specific interventions are expected to work (or not). The initial programme theory (IPT) considers the effectiveness or failure of assistive technology interventions for dyslexic adults. It explains why and how complex interventions such as assistive technology are adopted and/or abandoned in different contexts for and by adults with dyslexia. A co-design approach was adopted to develop a pilot questionnaire to collect data on assistive technology use and stigma experiences. Based on the pilot questionnaire and with the intent of co-design, the building of a standardised scale for multidimensional stigma was explored to assess the dimensions of assistive technology stigma for doctoral students with dyslexia.

3.1 Plan and Design: Phase One

To prevent further misunderstanding, a panel of international experts and doctoral students with dyslexia was established to ensure that expertise and lived experience stand on equal footing when exploring the stigma associated with assistive technology design and use. Among the members are experts in HCI, dyslexia, co-design, stigma research, and doctoral students with dyslexia with first-hand experience in navigating academia and utilising assistive technology. This approach established equality of collaboration in assistive technology design and addressed concerns that not all collaboration is perceived to be equal.

Part of the problem with dyslexia is its complexity, as addressed previously. A realistic review approach focuses on unpacking extensive evidence that distinguishes what interventions work or do not work, for whom, in what circumstances, and how and why (Wong et al. 2013). In this case, it meant uncovering the underlying assumptions about how assistive technology for adults with dyslexia is supposed to work (or not) and the expected impacts. Preliminary findings suggest a significant theme of perceived stigmatisation and discrimination in varied workplaces, which has been relatively underexplored in the empirical literature, specifically in HCI design.

Based on the initial realist review, a preliminary questionnaire was developed for collecting data and evaluating the effects of multidimensional stigmatisation on dyslexic academics. Panellists iteratively defined survey items and assessed the validity and reliability of the questionnaire instrument development. Survey results were used to identify how and why doctoral students use assistive technology in academia and to assess the potential effects of multidimensional stigmatisation, including assistive technology stigma.

Using the context-specific foundational data, the expert panel decided on ways to build a reliable and valid standardised instrument to measure multidimensional stigma for doctoral students with dyslexia. As of yet, there is no official scale for this. In using the self-reported survey, HCI designers can identify potential root causes of stigma to determine the effects on assistive technology use. This approach necessitated exploring prior work such as PAMS (Product Appraisal Model for Stigma), PIMS (Product Intervention Model for Stigma) and (PIADS) Psychosocial Impact of Assistive Devices Scale (Jutai and Day 2002; Vaes et al. 2016) and Frauenberger et al. (2015) "tool-to-think-with" that guides participatory design practices. In turn, using a multidimensional stigma scale for doctoral students with dyslexia may lead to more stigma-free academic research, improved assistive technology context interaction, and better stigma reduction efforts.

4 Discussion

The stigma attached to existing assistive technology use for adults with dyslexia is rarely talked about. The experience of stigma is subjective, shaped by lived experiences. Some designers may think that stigma is not likely to be harmful. Yet it is a silent influence on assistive technology adoption, and much remains to be learnt. With an enhanced co-design attitude and an understanding of real-world complexities and contextual constraints, universal design can better understand people with dyslexia hidden needs. In part, the collaboration seeks to make sense of multidimensional stigma, provide real-world insight, and generate new ideas from the perspective of those affected.

Although traditional design methods will be used to capture insights through written text, other co-design tools will also be explored in order to create new approaches to working with dyslexia data (Sanders and Stappers 2008). For example, even though passive methods such as interviews, self-reported surveys, and user observation can offer valuable insights, active creative techniques, such as design method cards, can also provide helpful insight into user experience (UX). Furthermore, combining ecological momentary assessments (EMA) with co-design practices can reduce the disadvantages of traditional design methods. Such approaches could improve survey response biases and improve progress in assistive technology design. Despite its complex nature, early-stage to end-stage collaboration may result in better understanding and prove effective in driving real change in dyslexia research. In general, "fuzzy front-end" (FFE) could be described as one of the most important considerations in design and execution (Sanders and Stappers 2008).

Amongst many things, Power and Petrie (2019) counsel against developing and validating standardised scales without including people with disabilities. In this case, by working together, assistive technology stigma-free design for academic workplaces may be better understood through validating scales with doctoral students with dyslexia. Future work will focus on effectively refining the role of standardised questionnaires in HCI practices for these specific users.

References

Crabb, M., Heron, M., Jones, R., Armstrong, M., Reid, H., et al.: Developing accessible services: understanding current knowledge and areas for future support. In: Proceedings of the 2019 Conference on Human Factors in Computing Systems, Glasgow, Scotland, UK, 4–9 May 2019 (2019). Paper: 216

Das, M., Tang, J., Ringland, K.E., Piper, A.M.: Towards accessible remote work: understanding work-from-home practices of neurodivergent professionals. Proc. ACM Hum.-Comput. Interact. **5**(CSCW1) (2021). Article 183. https://doi.org/10.1145/3449282

Desmond, D., Layton, N., Bentley, J., Boot, F.H., Borg, J., et al.: Assistive technology and people: a position paper from the first global research, innovation and education on assistive technology (GREAT) summit. Disabil. Rehabil. Assist. Technol. **13**(5), 437–444 (2018). https://doi.org/10.1080/17483107.2018.1471169

Doyle, N.: Neurodiversity at work: a biopsychosocial model and the impact on working adults. Br. Med. Bull. **135**(1), 108–125 (2020). https://doi.org/10.1093/bmb/ldaa021

Elliott, J.: It's time to be scientific about dyslexia. Read. Res. Q. **55**, S61–S75 (2020). https://doi.org/10.1002/rrq.333

Frauenberger, C., Good, J., Fitzpatrick, G., Iversen, O.S.: In pursuit of rigour and accountability in participatory design. Int. J. Hum. Comput. Stud. **74**, 93–106 (2015). https://doi.org/10.1016/j.ijhcs.2014.09.004

Hamilton, J.: How neurodiversity is changing work. ITNow **62**(1), 56–57 (2020). https://doi.org/10.1093/itnow/bwaa026

Jacobs, L., Parke, A., Ziegler, F., Headleand, C., De Angeli, A.: Learning at school through to university: the educational experiences of students with dyslexia at one UK higher education institution. Disabil. Soc. **37**(4), 662–683 (2022). https://doi.org/10.1080/09687599.2020.1829553

Johansson, S., Gulliksen, J., Gustavsson, C.: Disability digital divide: the use of the internet, smartphones, computers and tablets among people with disabilities in Sweden. Univ. Access Inf. Soc. **20**(1), 105–120 (2021). https://doi.org/10.1007/s10209-020-00714-x

Jutai, J., Day, H.: Psychosocial impact of assistive devices scale (PIADS). Technol. Disabil. **14**, 107–111 (2002). https://doi.org/10.3233/TAD-2002-14305

Kinsella, M., Waduud, M.A., Biddlestone, J.: Dyslexic doctors, an observation on current United Kingdom practice. MedEdPublish, **6**(1), 60 (2017). https://doi.org/10.15694/mep.2017.000060

Kirby, P., Nation, K., Snowling, M., Whyte, W.: The problem of dyslexia: historical perspectives. Oxf. Rev. Educ. **46**(4), 409–413 (2020). https://doi.org/10.1080/03054985.2020.1770020

Livingston, E.M., Siegel, L.S., Ribary, U.: Developmental dyslexia: emotional impact and consequences. Aust. J. Learn. Difficulties **23**(2), 107–135 (2018). https://doi.org/10.1080/19404158.2018.1479975

Locke, R., Alexander, G., Mann, R., Kibble, S., Scallan, S.: Doctors with dyslexia: strategies and support. Clin. Teach. **14**(5), 355–359 (2017). https://doi.org/10.1111/tct.12578

Macdonald, S.J., Cosgrove, F.: Dyslexia and policing: understanding the impact that dyslexia has in the police service in England and Wales. Equality Divers. Inclusion Int. J. **38**(6), 634–651 (2019). https://doi.org/10.1108/EDI-11-2018-0218

Major, R., Tetley, J.: Effects of dyslexia on registered nurses in practice. Nurse Educ. Pract. **35**, 7–13 (2019). https://doi.org/10.1016/j.nepr.2018.12.012

Meyer, M.W., Norman, D.: Changing design education for the 21st Century. She Ji: J. Des. Econ. Innov. **6**(1), 13–49 (2020). https://doi.org/10.1016/j.sheji.2019.12.002

Moll, K., Snowling, M.J., Hulme, C.: Introduction to the special issue comorbidities between reading disorders and other developmental disorders. Sci. Stud. Read. **24**(1), 1–6 (2020). https://doi.org/10.1080/10888438.2019.1702045

Moojen, S.M.P., Gonçalves, H.A., Bassôa, A., Navas, A.L., de Jou, G., Miguel, E.S.: Adults with dyslexia: how can they achieve academic success despite impairments in basic reading and writing abilities? The role of text structure sensitivity as a compensatory skill. Ann. Dyslexia **70**(1), 115–140 (2020). https://doi.org/10.1007/s11881-020-00195-w

Motti, V.G.: Designing emerging technologies for and with neurodiverse users. In: Proceedings of the 37th International Conference on the Design of Communication, pp. 1–10 (2019)

Nalavany, B.A., Logan, J.M., Carawan, L.W.: The relationship between emotional experience with dyslexia and work self-efficacy among adults with dyslexia. Dyslexia **24**(1), 17–32 (2018). https://doi.org/10.1002/dys.1575

NDA: The 7 principles. National Disability Authority. universaldesign.ie/what-is-universal-design/the-7-principles/. Accessed 20 Jan 2023 (2020)

Norman, D.A., Stappers, P.J.: DesignX: complex sociotechnical systems. She Ji: J. Des. Econ. Innov. **1**(2), 83–106 (2015). https://doi.org/10.1016/j.sheji.2016.01.002

Norstedt, M.: Work and invisible disabilities: practices experiences and understandings of non-disclosure. Scand. J. Disabil. Res. **21**(1), 14–24 (2019)

O'Brolcháin, F., Gordijn, B.: Risks of stigmatisation resulting from assistive technologies for persons with autism spectrum disorder. Technologies **6**(1) (2018). https://doi.org/10.3390/technologies6010027

Pichiliani, T.C.P.B., Pizzolato, E.B.: A survey on the awareness of Brazilian web development community about cognitive accessibility. In: Proceedings of the 18th Brazilian Symposium on Human Factors in Computing Systems, pp. 1–11, October 2019

Power, C., Petrie, H.: Working with participants. In: Yesilada, Y., Harper, S. (eds.) Web Accessibility. HIS, pp. 153–168. Springer, London (2019). https://doi.org/10.1007/978-1-4471-7440-0_9

Protopapas, A.: Evolving concepts of dyslexia and their implications for research and remediation. Front. Psychol. **10**, 2873 (2019). https://doi.org/10.3389/fpsyg.2019.02873

Ramus, F.: Should there really be a 'Dyslexia debate'? Brain **137**(12), 3371–3374 (2014). https://doi.org/10.1093/brain/awu295

Sanders, E.B.N., Stappers, P.J.: Co-creation and the new landscapes of design. Co-Design **4**(1), 5–18 (2008). https://doi.org/10.1080/15710880701875068

Schröppel, T., Miehling, J., Wartzack, S.: The role of product development in the battle against product-related stigma – a literature review. J. Eng. Des. **32**(5), 247–270 (2021). https://doi.org/10.1080/09544828.2021.1879031

Shakespeare, T.: Disability Rights and Wrongs Revisited, 2nd edn. Routledge, London (2013)

Shaw, S., Anderson, J.: Doctors with dyslexia: a world of stigma, stonewalling and silence, still? MedEdPublish **6**, 29 (2017). https://doi.org/10.15694/mep.2017.000029

Shinohara, K., Wobbrock, J.O.: In the shadow of misperception: assistive technology use and social interactions. In: Proceedings of the SIGCHI Conference on Human Factors in Computing Systems, Vancouver, BC, Canada, May 2011, pp. 705–714 (2011). https://doi.org/10.1145/1978942.1979044

Snowling, M.J., Hulme, C., Nation, K.: Defining and understanding dyslexia: past, present and future. Oxf. Rev. Educ. **46**(4), 501–513 (2020). https://doi.org/10.1080/03054985.2020.1765756

Syma, C.: Invisible disabilities: perceptions and barriers to reasonable accommodations in the workplace. Libr. Manag. **40**(1/2), 113–120 (2019). https://doi.org/10.1108/lm-10-2017-0101

Vaes, K., Stappers, P.J., Standaert, A.: Measuring product-related stigma in design. In: DRS 2016 – Future Focused Thinking. DRS Biennial Conference Series. Design Research Society. DRS Digital Library (2016)

Wong, G., Greenhalgh, T., Westhorp, G., Buckingham, J., et al.: RAMESES publication standards: realist syntheses. BMC Med. **11**(1), 21 (2013). https://doi.org/10.1186/1741-7015-11-21

Ymous, A.E.A.: "I am just terrified of my future"—epistemic violence in disability related technology research. In: CHI EA 2020, Honolulu, HI, USA, pp. 1–16, April 2020. https://doi.org/10.1145/3334480.3381828

Tinkerable Augmentative and Alternative Communication for Users and Researchers

B. Yang(✉) and P. O. Kristensson

Engineering Design Centre, Department of Engineering, University of Cambridge, Cambridge, UK
{by266,pok21}@cam.ac.uk

Abstract. Augmentative and alternative communication (AAC) users are highly diverse as their communication needs depend on many factors, such as motor ability, cognition, age, education level, and overall preferences. This chapter introduces the concept of *tinkerability* for AAC. We conjecture that *Tinkerable AAC (TAAC)* has the potential to empower AAC users, developers, and researchers to fine-tune, adapt, and explore better communication solutions *in tandem* with the system. We present a framework for tinkerable AAC that concretely links its qualities to AAC research and development areas and subareas. To demonstrate the viability of tinkerable AAC, we present a tinkerable predictive text entry system for it. It allows users to adjust many aspects of the system during use, ranging from the text prediction algorithms in the back end to the parameters underpinning the keyboard layout in the front end. In addition, it also allows AAC system developers to adopt state-of-the-art language models and enables AAC researchers to carry out *in-situ* research experiments. We discuss the design space opened up by tinkerable AAC and its potential in empowering the AAC community to co-design individually tailored solutions for users.

1 Introduction

Tinkering is an exploratory and iterative style of working where a user engages with a problem or a project. It is regarded as the opposite of *planning*, a more organised activity that is more direct but not repetitious (Resnick and Rosenbaum 2013).

Tailoring is an activity that modifies a system to fit specific use situations. It is distinct from *use*, as *use* refers to modifications made to the subject matter of the tool while *tailoring* refers to modifications of the tool itself. While *use* can give rise to an immediate perceived effect, the effect of *tailoring* is felt at a later stage (Henderson and Kyng 1995).

In this paper we suggest *tinkerability* to be the ability of a system to allow modification during interaction to fit different use scenarios and conditions required to reach specific goals, such as gaining higher efficiency. The modifications are not necessarily executed by the user directly as we envision that they can also be achieved by the assistance of AAC specialists, developers, and researchers.

We see many reasons why thinking about tinkerability in AAC is a promising research direction. First, AAC users form a very heterogeneous user group because individual

J. Goodman-Deane et al. (Eds.): CWUAAT 2023, *Design for Sustainable Inclusion*, pp. 137–145, 2023.
https://doi.org/10.1007/978-3-031-28528-8_15

preferences, capabilities, needs, and resources vary substantially. Allowing AAC users, specialists, and researchers to tinker with AAC systems provides opportunities to fine-tune systems for individuals.

Second, despite decades of research, AAC solutions are still suboptimal. Tinkerability encourages exploration and may serve two purposes: (1) increasing people's interest in improving AAC systems by explicitly observing the capabilities of sophisticated AAC systems; and (2) assisting AAC designers, developers, and researchers to elicit user requirements by empowering these users to further understand the capabilities and limitations of the systems they rely on.

Third, AAC systems relying on natural language processing and machine learning are becoming increasingly complex. Tinkerable AAC may help AAC designers and researchers to understand the increasing design space of AAC systems that has emerged due to the tremendous progress in artificial intelligence. By letting AAC designers experiment with AAC systems, tinkerable AAC allows them to observe potential benefits from such advancements first hand, thereby potentially to empowering both AAC users and professionals and stimulating new ideas and new formulations of design problems.

Thus, tinkerable AAC is a potential enabling technology, allowing non-experts in AI to create new futures within a broader AAC design space that encompasses state-of-the-art AI.

In the remaining space of this paper we will further explain tinkerability in AAC, present a framework for it, and exemplify it by presenting a tinkerable text prediction AAC system we have built for this purpose.

2 Tinkerability in AAC

AAC technology is used to assist people with communication disabilities to develop or regain their competence. Many computer-based AAC systems are developed using numerous representations for vocabulary concepts, including photographs, symbols, written words, letters of the alphabet, and so on (Beukelman and Light 2020). Due to users' varied physical capabilities, these representations can be accessed by fingers, toes, elbows, eyes, or use of multimodal approaches.

In the past decades, social inclusion has greatly increased (Mirenda 2014). As a result, literate nonspeaking individuals with motor disabilities tend to adopt a letter-by-letter spelling strategy to communicate, as it not only provides a more precise means of expression than images or symbols, but also brings a much wider range of expressions than preset text (Beukelman and Light 2020). Text prediction, including word prediction (Vertanen and Kristensson 2011), phrase prediction (McKillop 2018), sentence retrieval (Kristensson et al. 2020), and sentence generation (Shen et al. 2022) have been used to augment the language composition function of the AAC system, where vocabulary selection and message management are key points of system interaction.

Moreover, some AAC users, such as amyotrophic lateral sclerosis (ALS) patients, have progressive diseases which means their physical capabilities, and thus AAC needs, may change over time. For example, an ALS patient may originally type using a physical keyboard and later rely on eye gaze to communicate using an eye-typing system. Switching from one particular AAC system to a new one incurs upfront learning costs for both the user and the system: the latter likely needs to be configured and adapted to the user. This can be frustrating and challenging for AAC users, in particular, as AAC systems tend to be presented as black boxes. While tinkerability will not remove any costs in switching systems, it may help reduce frustration by allowing exploration of a new system to best fit new needs and wants.

In terms of what can be tinkered with, a computer-based system with a user interface (UI) can be partitioned into two components: the front end—what the user interacts with, and the back end—necessary system logic, such as prediction algorithms and language models. A tinkerable AAC system allows users to manipulate both the front and back end to meet their requirements and satisfy their curiosity as they explore the design space of the system.

However, even though a tinkerable AAC system would be beneficial, it remains difficult to develop one. Prior work (Henderson and Kyng 1995; Ellis et al. 2021) identifies four key points about developing a general tinkerable system. First, it needs to balance complexity and tinkerability, as tinkering requires additional buttons, switches, and mechanisms for adding behavior to code. Second, it becomes necessary to provide mechanisms for allowing design while in use. Third, the system must support saving and reestablishing the state of the system. Fourth, the three prior points all increase resource requirements in terms of time, money, experience, etc. to provide a working robust tinkerable AAC system.

There is also the challenge that the sheer complexity of state-of-the-art natural language processing systems means that the underpinning AI may be difficult to understand for users, designers, researchers, and developers alike. To allow a target audience with widely different educational backgrounds and interests to meaningfully tinker with a state-of-the-art word prediction and sentence prediction system, supporting sentence retrieval and sentence generation, requires careful considerations of which parameters to expose, how resulting system behavior can be analysed and communicated to end-users, and how sometimes perplexing AI behavior can be explained to users.

3 Tinkerable AAC Framework

To illustrate how a TAAC system contributes to AAC research and development, we propose a framework (Fig. 1) that links areas of AAC research and development to the qualities of a TAAC system. These links are indicated as purple arrows in the figure. Note that for clarity not all possible links are shown.

The framework divides the qualities of a TAAC into three areas: (1) input devices and techniques; (2) context sensing and AI; and (3) interaction parameters. Some of these qualities are easier to change than others. For example, to change the input device from a joystick to an eye-tracker necessitates either a very versatile TAAC hardware platform or an explicit change of hardware device. In contrast, changing the number of word suggestions shown in a word prediction display normally only involves simply changing a software parameter and ensuring there is space in the UI for an additional word suggestion.

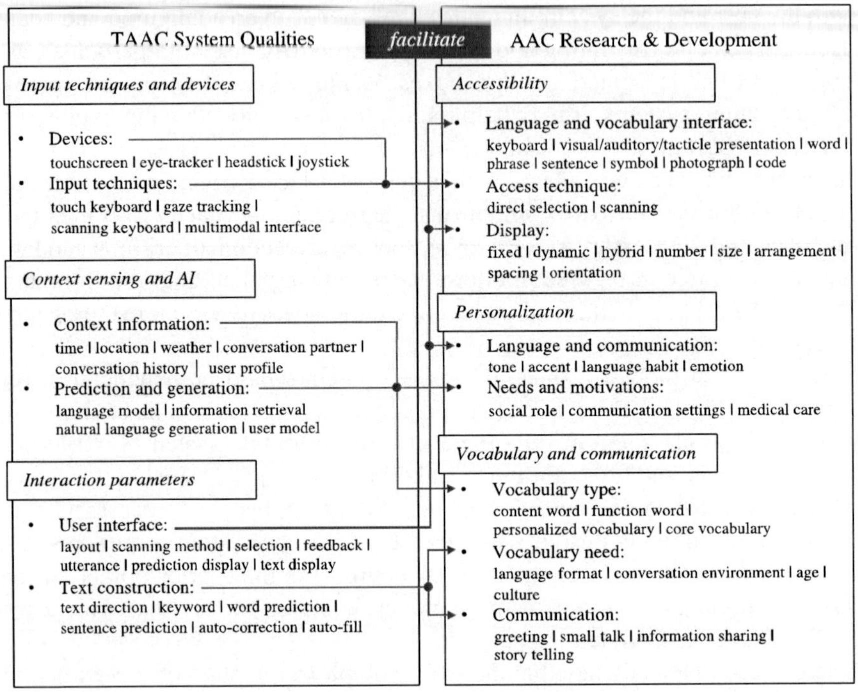

Fig. 1. A conceptual framework for explaining how qualities of a TAAC system (left) facilitate AAC research and development (right). The purple arrows indicate pathways linking TAAC system qualities that can be used for investigating related areas in AAC research and development. There are many pathways but we only show a few in the figure for clarity.

The framework also separates out AAC research and development areas. These are tentative, as a full scoping of this space necessitates a systematic literature review and interviews with researchers. For now, we identify three key areas: (1) accessibility; (2) personalisation; and (3) vocabulary and communication. We then break down each of these areas into subareas, covering research questions such as the expressiveness of the keyboard (under *Accessibility* → *Language and vocabulary interface*), and support for small talk (under *Vocabulary and communication* → *Communication*).

A way to use the framework is as a map for better understanding the design space of TAAC system qualities that can be explored for a particular area, or subarea, of interest. For example, tinkering with system utterances that reflect users' emotions can potentially benefit re-establishing the user's native language and further contribute to maintaining their social role, such as a family member. As another example, altering different language models and their underpinning parameters may improve language predictions in different communication settings.

The framework can also inform TAAC system design by enabling designers and researchers to work backwards from the research areas and subareas to ideate additional TAAC system qualities that can be incorporated into a TAAC system. In this way the framework effectively assists TAAC researchers to elicit requirements for a TAAC system that incorporates features with a direct research need.

Finally, in the long run we anticipate the use of systematic design engineering methods to make inroads in AAC as a complementary methodology (Kristensson et al. 2020). In this context, the TAAC framework can be used a map for AAC designers to understand which TAAC system qualities to investigate, for example, the number of generated sentence suggestions, to gain an understanding into appropriate areas of AAC research and development.

4 A Tinkerable AAC Text Prediction System

To demonstrate the viability of tinkerable AAC we have developed a tinkerable predictive AAC text entry system. This section introduces the design and features of this system and explains how it may assist AAC researchers and developers to iteratively improve the system. Figure 2 shows the system, including its keyboard, word predictions, and sentence predictions.

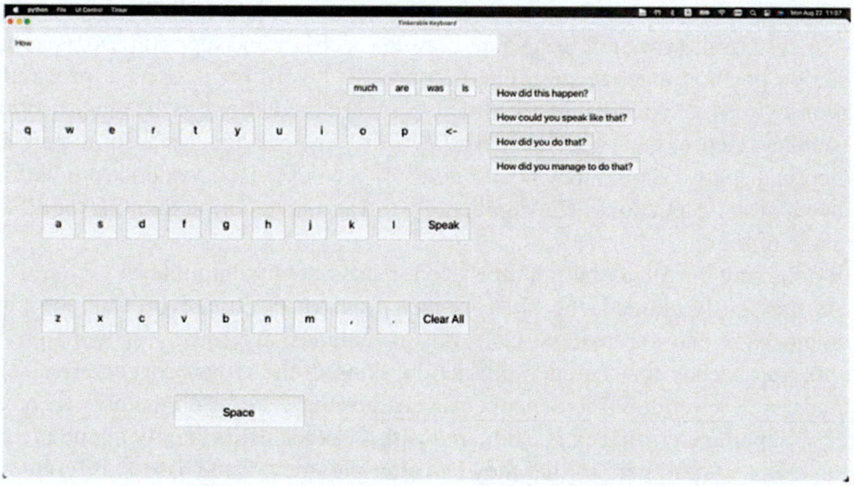

Fig. 2. The main user interface of the tinkerable AAC system

4.1 Interface

As shown in Fig. 2, the menu bar is separated into three sections: *File*, *UI Control*, and *Tinker*. *File* allows the user to save the current settings and reload previous settings as readable text files that can thus be edited outside the system. *UI Control* allows the user to move keys on the keyboard around by dragging them and thus changing the layout and geometry of the keyboard. Similarly, the position of predicted words and sentences can also be moved around by dragging them. Further system tinkering features associated with the back end are integrated into the *Tinker Panel* (see Fig. 3), which can be accessed by choosing *Tinker* in the menu.

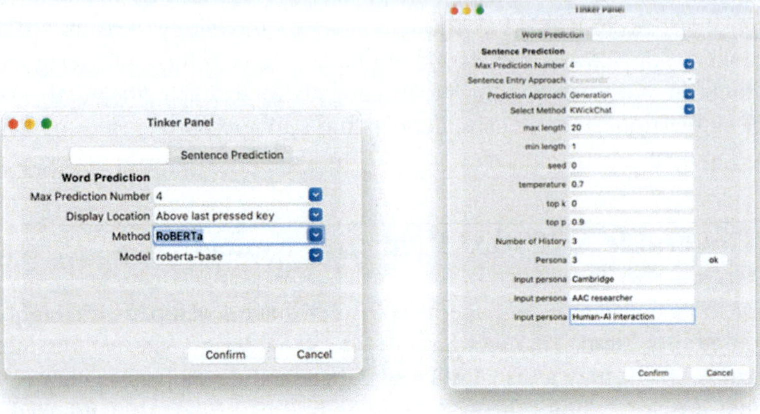

Fig. 3. The tinker panel for word prediction (left) and sentence prediction (right)

The system allows configuring gaps between each key and supports two types of word prediction position mechanisms. The first provides word predictions at prefixed static locations while the second dynamically places word predictions in the vicinity of the last pressed key. Hence, gaps between each keyboard row provide space for dynamic word prediction displays, should they be of interest to the AAC user. Sentence predictions can be shown in static locations. The number of word and sentence predictions are adjustable from one to four.

By clicking the *Speak* button the system reads out any output text.

As previously alluded, the *Tinker Panel* provides the user with an opportunity to configure back end parameters. Currently it includes five means of word prediction and six approaches for sentence prediction using both AI-based generative sentence methods and conventional sentence retrieval algorithms. For each method, there is a list of relevant parameters that can be tinkered with. This feature is mainly intended for AAC researchers and designers so that they can alter parameters and assess different models without needing to be able to program or to have any expertise in AI or information retrieval.

4.2 Text Prediction

Providing rich tinkerability around the central text prediction function has two main benefits. First, it allows customisation for individual users. A wide range of language prediction methods allows the AAC user or specialist to choose the most suitable method for the individual case. For example, some users require to send repeated commands (for example, "I need water" or "I am tired"). Such text retrieval can be achieved reliably using information retrieval methods, such as the approaches included by BM25 (Trotman et al. 2014) and SBERT (Reimers and Gurevych 2019).

For other users, communication can potentially be greatly speeded up by enabling users to provide keywords to prompt a language model to generate entire sentences, for example GPT-2 (Radford et al. 2019) and KWickChat (Shen et al. 2022).

These different methods have different tradeoffs and are hence complementary. For example, information retrieval methods may be beneficial in routine conversation with limited and repeated vocabulary, whereas sentence generation may be more efficient in open conversations, such as a casual chat where an exact utterance is not critical.

In addition, *how* to enable the AAC user to tell the system to generate or predict a sentence matters. Fundamentally, there are two approaches: either successive input of a series of words or the input of a series of keywords that serve as a prompt for a sentence generation algorithm. The two methods require a different interaction style, and one method may be preferred or more suitable for an individual user, or for a specific interaction context, such as a different environment or situation.

4.3 System Architecture

The tinkerable AAC system uses the model-view-controller (MVC) system architecture. MVC is one of the most widely used design patterns in software engineering of graphical user interfaces as it is simple and effective: it consists of three parts. In this system, the *model* corresponds to the back end, the algorithms and language models, the *view* is the user interface, including the keyboard (the front end), and the *controller* allows the user to interact with the front and back ends. This modular design facilitates iterative development of the system. For example, if an AAC user is used to inputting letters via a touchscreen but suddenly requires to use an eye-tracker, then an AAC developer merely needs to add an eye-tracker module to the view and link it to the controller, rather than developing a completely new system. Thereby the information retrieval and AI-based sentence generation subsystems, which require considerable development expertise, are reusable for a wide variety of different software development objectives.

We also conjecture that another advantage of this architecture might be that it simplifies integration of AAC systems that go beyond direct text entry. Many AAC researchers are interested in supporting conversations that are scenario-based, such as telling stories (Reiter et al. 2009), giving instructions (Todman et al. 2008), chat (Shen et al. 2022), and so on. With a flexible system architecture, developers can more easily integrate state-of-the-art language models into the system as a new UI option for AAC researchers, with minimal changes in the front end, and thereby benefit from the set of existing subsystems and algorithms that are already provided.

5 Discussion and Conclusions

This paper has introduced Tinkerable AAC (TAAC) and exemplified it with a tinkerable AAC system. While this system is a start towards a versatile enabling technology for user-involved AAC co-design, there are many improvements that are possible and many design considerations that need to be investigated.

First, from an end-user perspective any manipulation of front end aspects, such as the position of keys on the keyword and word - and sentence prediction slots, needs to be easy to understand and access. Further, any such manipulation needs to be easily reversible by the user.

Second, back end aspects, such as parameter choices and subsystem decisions need to be thoroughly explained. It will also be necessary to provide users, including designers, researchers, and developers, with easy methods to assess the efficacy of various parameter choices and activated subsystems. Ideally, a systematic approach is followed which could potentially be supported by the system by, for example, having workflows in the system for exploring and assessing parameter choices as a function of usage. This would enable designers and researchers to explore *what if* scenarios in a systematic way.

Third, an important objective of TAAC is to enable lightweight *in-situ* experiments that do not demand substantial development effort from researchers. Thus, it will be critical to provide easy-to-use and understand interfaces for setting up experiments and for logging and analysing data.

Fourth, it will be important to systematically evaluate any TAAC both in terms of benefits to end-users and in terms of the experience of designers, researchers, and developers.

In addition, the TAAC framework is a starting point for linking TAAC system qualities to AAC research and development areas and subareas. Further work is required, such as a systematic literature review and interviews with AAC researchers, to tease out additional research areas, and expand and refine them in the framework. With a more complete TAAC framework it is possible to work backwards from the research areas to the TAAC system qualities to identify gaps in support for research in a TAAC system.

In summary, we see TAAC as a promising direction in AAC but many challenges remain for it to truly take the form of an enabling technology democratising AAC design using state-of-the-art AI technologies. As examples of what is already possible, our existing system provides the scaffolding to assist with the following investigations: (1) assessing the impact of different text prediction layouts on text entry rate, cognitive load, and physical load; (2) understanding the impact of different language models on the AAC user experience; (3) analysing how language behavior changes when AAC users are in different environments; and (4) assessing the effect of context-aware text entry on communication rate. We hope TAAC will serve as a platform for incorporating recent AI advances for the benefit of AAC research and thus function as a catalyst for advanced AAC design exploiting state-of-the-art AI to ultimately improve the communication rate and user experience of AAC systems.

Open Science. Complete source code for the Tinkerable Keyboard can be found here: doi: https://doi.org/10.17863/CAM.91650.

References

Beukelman, D.R., Light, J.C.: Augmentative and Alternative Communication: Supporting Children and Adults with Complex Communication Needs. Brookes, Baltimore (2020)

Ellis, K., Dao, E., Smith, O., Lindsay, S., Olivier, P.: TapeBlocks: a making toolkit for people living with intellectual disabilities. In: Proceedings of the 2021 CHI Conference on Human Factors in Computing Systems, Yokohama, Japan, 8–13 May 2021, pp. 1–12 (2021)

Henderson, A., Kyng, M.: There's no place like home: continuing design in use. In: Baecker, R.M., Grudin, J., Buxton, W.A.S., Greenberg, S. (eds.) Readings in Human-Computer Interaction, pp. 793–803. Morgan Kaufmann (1995)

Kristensson, P.O., Lilley, J., Black, R., Waller, A.: A design engineering approach for quantitatively exploring context-aware sentence retrieval for nonspeaking individuals with motor disabilities. In: Proceedings of the CHI 2020, HI, USA, 25–30 April 2020, pp. 1–11 (2020)

McKillop, C.: Designing a context aware AAC solution. In: Proceedings of the 20th International ACM SIGACCESS Conference on Computers and Accessibility, Galway, Ireland, 22–24 October 2018, pp. 468–470 (2018)

Mirenda, P.: Revisiting the mosaic of supports required for including people with severe intellectual or developmental disabilities in their communities. Augment. Altern. Commun. **30**(1), 19–27 (2014)

Radford, A., Wu, J., Child, R., Luan, D., Amodei, D., et al.: Language models are unsupervised multitask learners. OpenAI Blog 9 (2019)

Reimers, N., Gurevych, I.: Sentence-BERT: sentence embeddings using Siamese BERT-networks. arXiv preprint arXiv:1908.10084 (2019)

Reiter, E., Turner, R., Alm, N., Black, R., Dempster, M., et al.: Understanding the storytelling of older adults for AAC system design. In: In: Proceedings of the 12th European Workshop on Natural Language Generation (ENLG 2009), Athens, Greece, 30–31 March 2009, pp. 1–8 (2009)

Resnick, M., Rosenbaum, E.: Designing for tinkerability. In: Honey, M. (eds.) Design, Make, Play, pp. 163–181. Routledge (2013). https://doi.org/10.4324/9780203108352

Shen, J., Yang, B., Dudley, J.J., Kristensson, P.O.: KWickChat: a multi-turn dialogue system for AAC using context-aware sentence generation by bag-of-keywords. In: Proceedings of 27th International Conference on Intelligent User Interfaces. Helsinki, Finland, 22–25 March 2022, pp. 853–867 (2022)

Todman, J., Alm, N., Higginbotham, J., File, P.: Whole utterance approaches in AAC. Augment. Altern. Commun. **24**(3), 235–254 (2008)

Trotman, A., Puurula, A., Burgess, B.: Improvements to BM25 and language models examined. In: Proceedings of the 2014 Australasian Document Computing Symposium, Melbourne, VIC, Australia, 27–28 November 2014, pp. 58–65 (2014)

Vertanen, K., Kristensson, P.O.: The imagination of crowds: conversational AAC language modeling using crowdsourcing and large data sources. In: Proceedings of the 2011 Conference on Empirical Methods in Natural Language Processing (EMNLP 2011), Edinburgh, UK, 27–29 July 2011, pp. 700–711 (2011). https://doi.org/10.5555/2145432.2145514

Towards a Multidisciplinary Approach for Designing Multimodal Sensory Communication Devices for Aeronautics

C. Lounis[1], M. Bénéjean[2], F. Hémont[2], P. Marrast[3], A. M. Brock[4](✉), H. Corderoy Du Tiers[5], J. L. Madras[5], and E. Ceretto[5]

[1] SII Research & ISAE-SUPAERO, Université de Toulouse, Toulouse, France
Christophe.LOUNIS@isae-supaero.fr
[2] University Rennes 2 – PREFICS, Rennes, France
{marie.benejean,florian.hemont}@univ-rennes2.fr
[3] CERTOP, UMR 5044, Université de Toulouse, Toulouse, France
philippe.marrast@iut-tarbes.fr
[4] ENAC, Université Toulouse, Toulouse, France
anke.brock@enac.fr
[5] Fans4All Association, Paris, France
contact@fans4all.org

Abstract. Deaf pilots in France are currently allowed to fly planes with the help of a second pilot handling voice and radio communication. Yet they are not allowed to pilot independently. Fans4All is an association that aims at making aeronautics more accessible to pilots who are hearing or speech impaired (HSI). In this paper we present our experience as a multidisciplinary design team (including two HSI pilots) working towards this goal. We present current and past steps to develop a Multimodal Sensory Communication Device (MSCD) composed of a touchscreen tablet and a haptic jacket, as well as the visual vocabulary to define messages between HSI pilots and air traffic controllers. Moreover, we present our approach combining quantitative and qualitative methods for evaluation. We hope that our work will help to make aeronautics more accessible to people with impairments.

1 Introduction

In a controlled airspace, vocal interaction between pilots and air traffic control (ATC) is mandatory for safety reasons. This interaction encompasses instructions from ATC to pilots as well as requests from pilots to ATC. This interaction implements a vocal language based on short sentences whose glossary and grammar are internationally defined, for example: aircraft: "ready for landing", ATC: "clear to land wind 160 [direction] 8 knots [force]". As an example, during the departure and arrival phases which each last around 10 min, pilots and ATC exchange around five interactions.

Commercial aviation pilots and air traffic controllers pass severe tests regarding their physical and psychological aptitude (including hearing). General Aviation however is less restrictive. Regulations vary between different countries, but generally, there are limitations regarding piloting for people with impairments. In France, light aviation pilots

J. Goodman-Deane et al. (Eds.): CWUAAT 2023, *Design for Sustainable Inclusion*, pp. 146–155, 2023.
https://doi.org/10.1007/978-3-031-28528-8_16

with hearing or speech impairments (HSI[1]) are currently allowed to fly in controlled airspace with a second pilot handling voice communication with ATC. They developed an alternative communication strategy where the second pilot transcribes audio messages on a whiteboard. Yet HSI pilots are not allowed to fly independently. However, an innovative silent interaction between an HSI pilot and ATC could be achieved through the use of digital media.

Sensory substitution devices offer a way for a person with impairments to overcome their disability. The sensory experience that these devices allow is not that of degraded perception, but of a new experience. The loss of one sensory function can be compensated for conveying missing information through an intact sense (Kristjánsson et al. 2016). Safety, Usability and User Experience are crucial concepts when designing new devices especially in the aeronautical context. Pilots face specific challenges from the need to control complex systems in a multitask environment which requires prediction in an uncertain world. The inter-related concepts of situation awareness and workload are central to aviation psychology. Three components of situation awareness are spatial awareness, system awareness, and task awareness (Endsley 1988). Task management is directly related to mental workload, as the competing demands of tasks for attention exceed the operator's limited resources. When designing new systems special attention should be paid to avoid loosening this element.

FANS4all (future air navigation systems for all) is an association created in 2017 to support French HSI pilots to be included by the French Civil Aviation Authorities (DGAC). In collaboration with its research partners FANS4all aims to define and implement a solution that allows HSI pilots to fly independently. This collaboration is multidisciplinary, combining engineering sciences with human and social sciences. It is supported by several public and private partner institutions, and several piloting clubs. HSI pilots are involved in the design process as end users and co-designers.

Figure 1 shows an overview of the different activities of the project. Following situated and distributed approaches to communication, cognition, and action (Hutchins and Klausen 1996; Theureau 2010), we analyse the piloting activity in the more global activity system to which it belongs and of which ATC is part. The development of a Multimodal Sensory Communication Device (MCSD) should allow the HSI pilot and air traffic controller pair to co-construct the meaning of the ground-aircraft messages in an efficient and safe way for the air operations required in the context of flight. We are particularly interested in shared sensemaking to identify the meanings constructed in situ (Weick et al. 2005) and the modalities of action that develop in the context of deafness. In our project we use complementary qualitative approaches (auto-confrontation) and quantitative approaches. In this paper, we present the part of the Fans4All project which we have worked on until now (marked in orange in Fig. 1) and our methodology for enabling more accessibility in aeronautics.

2 Related Work

Deaf people largely rely on the use of sign language for spoken information. One way to support this through technology is using sign language datasets (Bragg et al. 2021). An

[1] We use the term HSI instead of DHH (deaf or hard of hearing) because FANS4all project is also directed at people/pilots with speech difficulties.

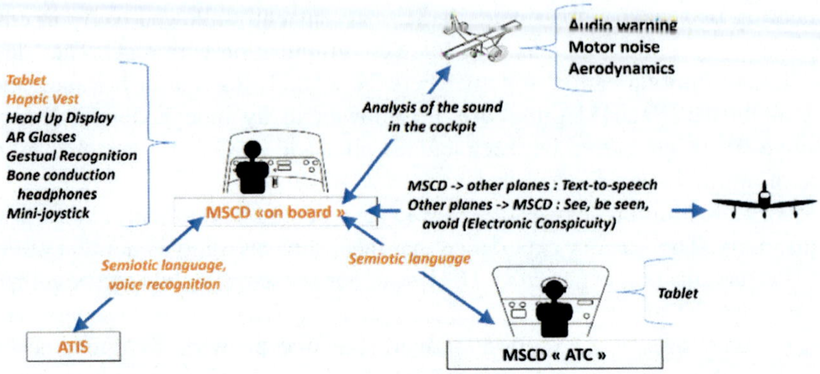

Fig. 1. Overview of the Fans4All project. Parts of the project that have already been addressed and are presented in this paper are highlighted in orange.

alternative is the generation of sign language avatars (Brock et al. 2020). Yet not all HSI people speak and understand sign language. Nonverbal sounds, such as environmental noise, can be presented to HSI users through visual or tactile modalities. For instance, sound has been represented on a wrist-worn tactile prototype (Jain et al. 2020). Recently, Jain et al. (2021) have proposed a taxonomy for making nonverbal sound in virtual reality accessible to deaf and hard of hearing users.

To our knowledge, few works have explored accessibility in aeronautics. Blind pilots in France use the soundflyer, a sonification system which sonifies two dimensions of the aircraft attitude, i.e., pitch and bank angles. This sonification consists in modulating the features (i.e., frequency, rhythm, inter-aural balance) of a sinusoidal pure tone which is continuously displayed to the pilot via his headphones (Valéry et al. 2017). In contrast to these technological advances for blind pilots, no system is used for HSI pilots to the best of our knowledge.

3 Design Methods

In this work, we apply a participatory design approach, involving end users in an iterative process. Fans4All currently has approximately 15 HSI members. Two HSI pilots are closely involved throughout the design process and hence co-authors of this paper. Moreover, we involved flight instructors, and Air Traffic Controllers, as well as members of the DGAC, and the design team, during different workshop from design to flight simulator assessments.

FANS4all applies a design approach using four successive steps. The four steps are gradually increasing use of sensory substitution and consequently autonomy of pilots, with the goal for a fully independent flight in 2030. Each step includes laboratory work, user studies in a simulator and finally in a real plane. To study the piloting activity, we developed with the support of pilots a set of scenarios which treat a part of the flight situations which the pilots meet most frequently: e.g., runway pattern, exit or integration of air zones, or breakdown. These scenarios then serve as a basis to compose flights in flight simulators by varying the complexity of the flight experience.

Fig. 2. Pilot in the flight simulator with the MSCD (tablet attached to the knee and haptic vest) and EEG used for experimental studies

3.1 User-Centered Multimodal Sensory Communication Device (MSCD) Design

The MSCD is composed of a tablet (Fig. 3) coupled with a haptic vest (Fig. 2) that directs the wearer towards the tablet when it vibrates. As part of the development of the MSCD, our objectives were 1/to create a visual transcription of verbal ATC messages (Text to Form protocol), 2/to create a way to compose messages. To do so we label the pilots' messages with keywords, thus allowing us to create buttons with short labels on the tablet.

Studies about the perception of signs are numerous, and sometimes contradictory. Some explain that reading a symbol is faster than reading a text (Velichkovsky et al. 1997), but some more recent works says that understanding a text is faster than understanding a symbol (Shinar and Vogelzang 2013). In any case, the latter survey suggests that reading speed increases with familiarity with symbols (a symbol proceeds from a semiotic break with what it represents (Bougnoux 2001), and in fact requires a learning phase). Moreover, lab studies generally do not reproduce factors, such as turbulence or noise which undoubtedly modify the capacity to apprehend texts.

Regarding the co-production of scenarios, we first identified the corresponding aeronautical phraseology. In a second step, we co-developed (with hearing pilots, HSI pilots, and ATC) a visual representation of this phraseology.

An iterative method was used for developing a catalogue of symbols.

Concerning the creation of buttons, the card sorting method (Wood and Wood 2008) was used to organise and categorise pilots' knowledge. An optimal information architecture was selected by the pilots during a participatory workshop. Finally, regarding the labels, an online MCQ was used a few weeks later to check correct understanding of the labels.

A bottom-up information architecture was developed respecting the visual logic of the cockpit. Consequently, the flight phases are located at the bottom of the interface and the sub-phases are located on top. We also labeled the buttons with keywords that pilots are familiar with. Pre-filled buttons divided into two columns allow pilots to send requests to ATC or allow readback (repeating what a controller says to indicate that the pilot has understood what he asked). Moreover, a chat allows to display the discussion history between the controller and the pilot. The header provides emergency buttons. Finally, Automatic Terminal Information Service (ATIS) can be consulted at any time during a flight. ATIS is a vocal message containing essential information, such as weather, active runways, available approaches and any other information needed by pilots. Pilots usually listen to ATIS before contacting control, which reduces ATC workload and decreases frequency occupancy. Yet since this is an audio-based service it is currently not accessible to HSI pilots. We developed an accessible ATIS application which has been presented in (Brock et al. 2022).

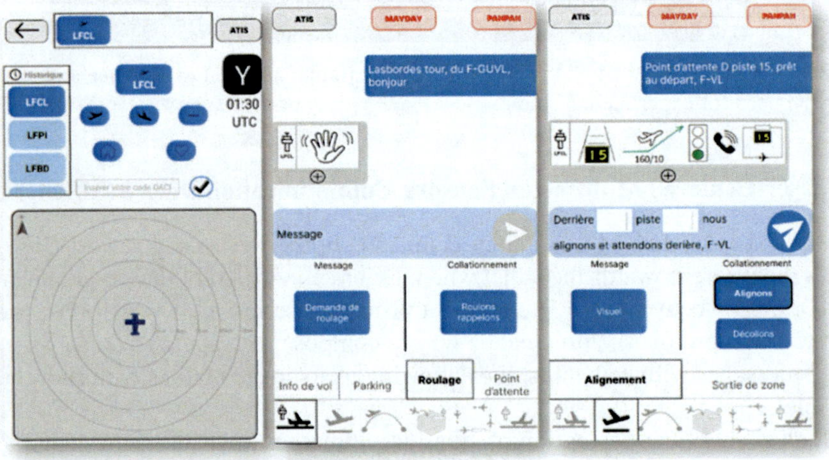

Fig. 3. MSCD tablet interface (left) ATIS application, (middle) taxiing interface, (right) alignment on the runway

To create haptic salience and haptic messages, we made use of "Tactons" which are vibro-tactile messages defined through frequency, amplitude, waveform and duration of a tactile pulse, plus body location (Brewster and Brown 2004). We used a haptic vest with 2×20 vibrating motors in the back and front. We implemented a pattern to attract attention which consists of a sequence of 2 vibrations of 1 s each, with all front and rear motors at 80% intensity (medium feel) at a 200 Hz constant vibration frequency. The pattern for throttle-up message consists of an 8 s sequence of 5 similar vibrations of 1,5 s each, with all front and rear motors at 100% intensity (maximum feel). The repetitive vibratory pattern starts from the center of the back, widens outwards, resumes on the outside of the front face and ends in the center at the front.

3.2 Methodology for Simulator Studies

Flight simulator experiments were conducted to evaluate the MSCD design. A set of elements were collected through diverse sensors (flight parameters, EEG & ECG data, and various video recordings such as 360° camera, tablet screen capture, etc.). Each simulator session was composed of flights with the MSCD prototypes and control flights with a radio copilot as used in real piloting activities. At the end of these simulations, we proceeded to survey battery assessing subjective Workload with NASA TLX (Hart and Staveland 1988), subjective Situation Awareness with SART (Selcon and Taylor 1990), Usability and User-experience respectively with SUS (Brooke 1996) and CSUQ (Lewis 1995) to obtain subjective quantitative indicators (summary in Table 1).

We also led self-confrontation through a re-enactment using material traces of the activity (Theureau 2010) assisted by video replay of the flight to co-produce qualitative feedback with the pilots. We ended the process with a semi-directive interview accompanied by wireframes to record feedback on the modification of the interfaces (Figs. 4 and 5). This methodological assembly allows us to cross a macro vision of the flight experience with more precise elements and moments of this same flight.

Fig. 4. Mosaic showing self-confrontation method and material

Fig. 5. Photography showing a self-confrontation session

Table 1. Experimental Protocol with different conditions and measurements

	Quantitative	Qualitative
Objective	ECG	SALIANT
	EEG	
	Flight parameters	
Subjective	NASA TLX	Self-confrontation
	SART	
	SUS	Semi-directive interview
	CSUQ	

Caption
Workload
Situation Awareness
Usability
User-Experience
Sense-Making

4 Results of Analysis from the Self-confrontation Interviews

Five principal dimensions of analysis have emerged from the self-confrontation:

1. **Communication, an additional workload:** For HSI pilots, the task of communication appears to be an additional workload during the management of their flight, which they previously delegated to the radio pilot. This element of analysis raises questions about the training for this communication work and for the used device. A training device is planned to be developed once the MSCD is functional.
2. **Vision, a sense that should not be overloaded:** VFR ("visual flight rule") expects to lose as little as possible visual contact with the aircraft's environment. As the tablet was attached to the pilots' thighs, simulation tests show deviations, altitude variations, etc. of the flight trajectory while pilots were reading or composing a message. It is also necessary to think about the trajectory of the gaze from tablet to flight instruments and what is happening outside, to facilitate the management of information and the manipulation of the communication and piloting tools.
3. **Communicational temporality and action:** Take-off and landing require particular attention and mobilise the pilot's body. The writing time on the tablet seems long compared to the actions expected in these phases of flight. We should not underestimate the work of "data entry".
4. **From reading the message to understanding the situation, a multidimensional work of meaning co-production:** The phraseological statements do not determine the meaning of the situation by themselves. This emerges from the context, i.e., from the interactions between the individuals and the material, social, linguistic, and cultural resources at their disposal. Concerning the interface of the tablet, we noticed that the possible answers proposed (and displayed on the screen) can cause inferences in the comprehension of the message which the pilot answers. The phraseological statement and its matched response collation must be considered in relation to the "question-answer" pair and other information sources that may confirm or refute understanding of the message and the situation.
5. **From situations to communicative perception:** The tests of the haptic messages were conclusive in terms of attentional attraction for all three pilots who tested it. The

pilots easily understood that they had to look at the tablet screen because an update of the displayed information required their attention. For a throttle-up message the results varied, from "I haven't felt the vibration", to "I totally understood". The approach phase requires much more attention from the pilots, who must control many parameters and intervene much more for trajectory control. Knowledge of the flight area, the number of hours of flying practice and experience of flying in a controlled area are all parameters that influenced the perception of the message and should be considered in the future.

5 Conclusion and Perspectives

This paper presents an overview of the current and past activities of the Fans4All project which aims to make aeronautics more inclusive for pilots who are hearing and speaking impaired. We present our user-centered and iterative design approach, using qualitative and quantitative approaches, as well as the developed MSCD prototype composed of a tablet and haptic vest. This methodological coupling allows us to combine general evaluations of each simulation (and to make comparisons between them) with a more detailed understanding of each of them.

In the future it will be interesting to include further devices in the MSCD, such as head-up displays, or Augmented Reality Glasses, as suggested in Fig. 1. We are currently also working on a tablet which will be used by ATC to communicate with the pilots' MSCD tablets.

We will also conduct further experimental studies with HSI and hearing pilots. Pilot studies with the tablet show incompatibilities (or at least complications) between reading/composing a message on a tablet and landing the plane while pilots need both hands to manipulate the plane. Perspectives include testing a small joystick fixed on the airplane's control stick to compose messages, integrating a screen situated in eyesight in the cockpit, or installing a portable screen device closer to the top of the cockpit.

Concerning the development of visual semiotics, the work on the "Text to Form" protocol will continue to progress as new scenarios are developed. We will test the capacity of this visual language to translate more and more complex messages and situations. Moreover, the next developments will consider sign language in its syntax, but also in its visual aspect.

The haptic vest which was used to attract attention and provide "throttle up" alert messages, has moreover shown its limits. In the future, we will consider adaptation to morphology, characteristics and distribution of motors and vibrations, and the unique haptic modality. Complementary approaches on the haptic part of the MSCD aim at exploring other vibratory patterns (notifications, phraseology elements, alerts) and other interaction modalities (bone conduction, kinesthetic rendering, other vibration modes).

With this project we hope to contribute to making aeronautics more accessible. The outcomes obtained during the Fans4All project for HSI pilots will moreover be applicable to hearing pilots of General Aviation (and maybe commercial aviation) who no longer meet the regulatory medical requirements for hearing due to ageing or hearing loss, to define a backup device that can be used when the radio link with ATC or ATIS is broken and to define a device that can improve situational awareness and reduce occurrences of « inattentional deafness».

References

Bougnoux, D.: Introduction aux sciences de la communication. La Découverte, Paris (2001)

Bragg, D., Caselli, N., Hochgesang, J.A., Huenerfauth, M., Katz-Hernandez, L., et al.: The FATE landscape of sign language AI datasets: an interdisciplinary perspective. ACM Trans. Access. Comput. **14**(2), 1–45 (2021). Article 7

Brewster, S.A., Brown, L.M.: Non-visual information display using tactons. In: CHI 2004 Extended Abstracts on Human Factors in Computing Systems, Vienna, Austria, 24–29 April 2004, pp. 787–788 (2004)

Brock, A., Classe, C., Duphil, L., Fabas, F., Guigo, L., et al.: Making ATIS accessible for pilots who are deaf or hard of hearing. In: ICCAS 2022 - International Conference on Cognitive Aircraft Systems, Toulouse, France, 1–2 June 2022 (2022)

Brock, H., Law, F., Nakadai, K., Nagashima, Y.: Learning three-dimensional skeleton data from sign language video. ACM Trans. Intell. Syst. Technol. **11**, 30 (2020)

Brooke, J.: SUS-a quick and dirty usability scale. In: Usability Evaluation in Industry, 1st edn. CRC Press (1996)

Endsley, M.R.: Design and evaluation for situation awareness enhancement. In: Proceedings of the Human Factors Society Annual Meeting, vol. 32, no. 2, pp. 97–101 (1988)

Hart, S.G., Staveland, L.E.: Development of NASA-TLX (task load index): results of empirical and theoretical research. In: Hancock, P.A., Meshkati, N. (eds.) Human Mental Workload, pp. 139–183. North-Holland (1988)

Hutchins, E., Klausen, T.: Distributed cognition in an airline cockpit. In: Engeström, Y., Middleton, D. (eds.) Cognition and Communication at Work, pp. 15–34. Cambridge University Press, Cambridge (1996)

Jain, D., Chiu, B., Goodman, S., Schmandt, C., Findlater, L., et al.: Field study of a tactile sound awareness device for deaf users. In: Proceedings of the 2020 International Symposium on Wearable Computers, Virtual Event Mexico, 12–16 September 2020, pp. 55–57 (2020)

Jain, D., Junuzovic, S., Ofek, E., Sinclair, M., Porter, J., et al.: A taxonomy of sounds in virtual reality. In: Designing Interactive Systems Conference 2021 (DIS 2021), pp. 160–170. Association for Computing Machinery, New York (2021)

Kristjánsson, Á., Moldoveanu, A., Jóhannesson, Ó.I., Balan, O., Spagnol, S., et al.: Designing sensory-substitution devices: principles, pitfalls and potential 1. Restorative Neurol. Neurosci. **34**, 769–787 (2016)

Lewis, J.R.: IBM computer usability satisfaction questionnaires: psychometric evaluation and instructions for use. Int. J. Hum.-Comput. Interact. **7**(1), 57–78 (1995)

Selcon, S.J., Taylor, R.M.: Evaluation of the situational awareness rating technique (SART) as a tool for aircrew systems design. In: AGARD, Situational Awareness in Aerospace Operations, 8 p (SEE N 90-28972 23-53) (1990)

Shinar, D., Vogelzang, M.: Comprehension of traffic signs with symbolic versus text displays. Transport. Res. F Traffic Psychol. Behav. **18**, 72–82 (2013)

Theureau, J.: Les entretiens d'autoconfrontation et de remise en situation par les traces matérielles et le programme de recherche «cours d'action». Revue D'anthropologie des Connaissances **4**, 287–322 (2010)

Valéry, B., Scannella, S., Peysakhovich, V., Barone, P., Causse, M.: Can an aircraft be piloted via sonification with an acceptable attentional cost? A comparison of blind and sighted pilots. Appl. Ergon. **62**, 227–236 (2017)

Velichkovsky, B., Sprenger, A., Unema, P.: Towards gaze-mediated interaction: collecting solutions of the "Midas touch problem." In: Howard, S., Hammond, J., Lindgaard, G. (eds.) Human-Computer Interaction INTERACT '97. ITIFIP, pp. 509–516. Springer, Boston (1997). https://doi.org/10.1007/978-0-387-35175-9_77

Weick, K.E., Sutcliffe, K.M., Obstfeld, D.: Organizing and the process of sensemaking. Organ. Sci. **16**, 409–421 (2005)

Wood, J.R., Wood, L.E.: Card sorting: current practices and beyond. J. Usability Stud. **4**(1), 1–6 (2008)

Design Methods

Exploring Generative Design for Assistive Devices

C. Y. Choi[1] (✉) and D. Rosen[2]

[1] Royal College of Art, Design Products, London, UK
christina.choi@rca.ac.uk
[2] Institute of High Performance Computing, A*STAR, Singapore, Singapore
rosendw@ihpc.a-star.edu.sg

Abstract. This paper discusses the potential of applying generative design to assistive devices. The concept of a cyber-physical human system is presented. An example case of generative design to explore innovative design solutions with novel features is presented. Outstanding challenges and gaps in research are discussed.

1 Introduction

Users of assistive devices represent a continuum of abilities (Cook and Polgar 2008), from those with slight to moderate disabilities with more general needs to those with more severe disabilities with unique and specific needs. An assistive device (defined by the federal Individuals with Disabilities Education Act of 1990) may therefore function well for one group of users but poorly for another group. Further adding to the challenge is that the capabilities will vary not only across. Different populations but also change for individuals as their needs evolve over time.

This makes the design of assistive devices to fit particular needs a special challenge. User centered methods that involve users at various stages of the design process are common to help get the right mix of features and functions as well as testing. These approaches are used for unique and specialized conditions as well as more general ones where a more universal solution appropriate for a wider variety of users can be found.

The process from beginning to final product can be slow to yield a product that must find success in a segmented, niche market. Further, as new technologies and innovations become available, they can be slow to be widely integrated into new designs. Innovations with large changes are riskier and can take more time to develop with unpredictable impact on user acceptance.

Assistive devices often then go through an elaborate but mature process of matching a user to an available device that meets their needs and capabilities. It generally follows eight main steps (Kairalla et al. 2016): Referral and Appointment; Prescription; Product Preparation; User Training; Assessment; Funding and Ordering; Fitting; Maintenance, Repairs and Follow-Up. There are also many actors involved at different stages, including providers (physicians, physical/occupational therapists), payer (public/private insurers, government), suppliers (device manufacturers, resellers), and the clients (patients, caregivers or employers).

J. Goodman-Deane et al. (Eds.): CWUAAT 2023, *Design for Sustainable Inclusion*, pp. 159–168, 2023.
https://doi.org/10.1007/978-3-031-28528-8_17

While the process seeks to find a good match, abandonment is still common. This can result from a functional match that is not ideal, to other complex issues of perceived stigma, aesthetics or other preferences (Sugawara et al. 2018; Dos Santos et al. 2022). All of the complexity leads to opportunities for improvements and efficiencies in the process of design and delivery, especially for transformative design innovations or incremental incorporation of new technologies such as smart devices, assistive robotics and novel human/machine interfaces (e.g. brain-computer interfaces, eye tracking and facial gesture recognition) (Dicianno et al. 2019), wearable computing devices, or "Internet of Things" (IoT) devices and applications (de Domingo 2013; Rosen and Choi 2021).

The rest of this paper discusses proposals for meeting some of the opportunities and challenges. In particular we explore the potential of leveraging generative design for novel, customized engineering solutions for assistive products based on its foundations in product service systems and cyber physical human systems (CPHS). We also discuss the significant challenges of maintaining the voice of the user in such systems to evaluate the likely usability and appropriateness of generated designs.

2 CPHS, Generative Design and Assistive Devices

The starting point here for generating design solutions lies in the field of Cyber-physical-social systems (CPSS). These are typically considered as an evolution of cyber-physical systems to include human interactions; further it is thought that interactions among humans through CPSS can lead to emergence of social or community structures and behaviors (Yilma 2021). We can identify a subset of CPSS as cyber-physical-human systems (CPHS) as smart products that offer services its customers, supported by back-end systems (e.g., information, finance) and possibly other infrastructure. That is, CPHS are scoped to the individual and their interactions with the physical and cyber systems. This emphasis on the individual is not meant to lessen the importance of the social and community aspects of CPSS, but rather to reduce the research scope to individual human interactions without the complications arising from social (human-human) interactions.

Generative design of CPHS refers to the generation of many alternative designs that designers or customers can consider during early design stages (Autodesk 2022). The term "generative design" takes its meaning from the current geometrical design generators that resemble topology optimization solvers (Bendsoe 1995). Applied to CPHS, generative design has an expanded meaning that generates product configurations and layouts; software, computation, and communications systems; and user interfaces and interactions.

"Human" in CPHS refers to the goal of designing user/customer experiences and user interfaces, ensuring user acceptance, and, overall, ensuring that users gain value from the CPHS. This is consistent with the objectives of product-service-system (PSS) design but, we believe, encompasses additional considerations. Service design can be included under user/customer experiences and value gained. The motivation is the idea that a common design methodology can be developed for CPHS including smart products, products through which extensive services are delivered, and even intelligent manufacturing systems.

Generative design is used currently for topology optimization based software that synthesizes part shapes given structural design requirements. The software explores the design space to generate many potential solutions through which designers can browse and select designs to engineer and optimize further. These software systems incorporate some manufacturing constraints or allow the designer to specify a target manufacturing process. Outputs from generative design software are one or more geometric part models.

We believe that a broader perspective on generative design enables the solution of a much wider range of design problems, specifically CPHS design. However, for CPHS design, or even the design of assemblies, a different foundation is needed, one that reasons about product architectures, including function-form relationships. Additionally, design entities (components) should denote software modules, service elements, physical components, modules, and other constituent elements that comprise CPHS.

One aspect of this broader perspective on generative design is the capability to generate design configurations. That is, configuration design encompasses the selection of constituent elements, their connections and logical and spatial relationships, and their hierarchical organization. While conventional optimization explores design spaces that are subsets of $\mathbf{R}^n$ (i.e., design variables are real-valued dimensions and attributes), configuration design operates in large combinatorial design spaces where the analogs of design variables are discrete choices of constituents or relationships between constituents. Each element of such design spaces is a design configuration, that is, a collection of elements with relationships that may represent a partial or complete design solution. These design spaces have been described and utilized for product family design in our earlier work (Siddique and Rosen 2001; Hansen and Rosen 2019). Since each design space element can have associated attributes and/or dimensions, each element represents its own continuous design space (subset of $\mathbf{R}^n$). As a consequence, configuration design spaces are large-scale mixed discrete-continuous spaces (called mixed-discrete).

To get a sense of the structure of such design spaces, consider the combinatorial space defined by selecting collections of constituents from a set of 5 components. That is, each design space element is a subset of the 5 components. This collection of sets of subsets can be arranged hierarchically in a subset-superset lattice as shown in Fig. 1. This type of hierarchy is a partially ordered set, ordered by the subset relationship. More generally, elements of design spaces can be considered as graphs, where the relationships between constituents are modeled as edges of a graph, and the ordering relation is by subgraph.

These combinatorial mixed-discrete design spaces are presented here to highlight the importance of defining the design space before developing methods to search for design solutions. Discrete mathematics, including set theory and graph theory, can be applied. Combinatorics can estimate design space size. Further, design requirements should be modeled such that feasible regions of the design space can be identified and distinguished from infeasible regions. Search and sampling methods should be guided by the underlying mathematical structure of the design space, and should avoid infeasible regions. Generative design of CPH systems incorporates much beyond configuration design as outlined here. But configuration design serves as a significant element of the larger design challenge.

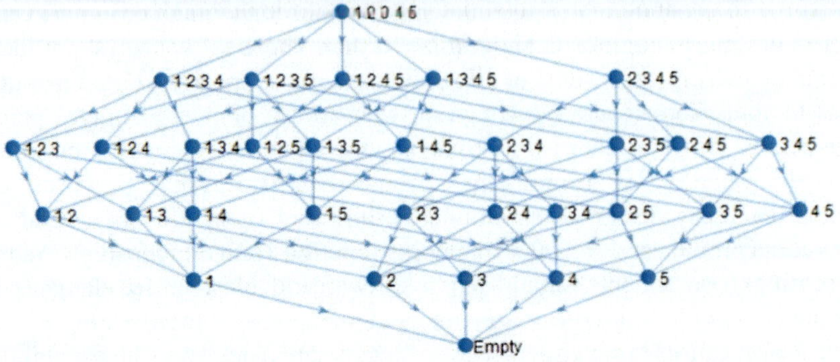

Fig. 1. Partially ordered set of all subsets of (Amrehn et al. 2019; Autodesk 2022; Bendsoe 1995; Brooke 1996; Brooke 2013)

These ideas can be applied to an example of a wheelchair (WC) CPHS family design. This starts with an identification of the highest level functions that the proposed WC family will perform. These include Transport Patient, Monitor Patient, Diagnose Patient, Monitor WC, and Maintain WC. Of these, the primary function that will be common to all CPHS's in the family is (of course) Transport Patient. An example view model for this function is shown in Fig. 2, with a top-level Receiver State Parameter (RSP, indicator of customer value) of Safe & Secure Mobility identified. Associated with each sub-function is either a more specific RSP or a Function Parameter (FP). To support design of a family of WC CPHS's, usage scenarios have been developed for WC maintenance, user health monitoring, health interventions, etc. with resources identified to provide requirements for the future WC CPHS family.

At present, companies offer services to select and fit a WC to a patient, maintain the WC, and upgrade it with new capabilities. These services are provided through phone calls, websites, and technician visits. By adding sensors to the WC, we believe that additional services can be provided. Furthermore, these services will be delivered through websites with considerable back-end information technology (IT) that collects sensor data, analyzes it, and notifies appropriate personnel when the need arises. The back-end IT infrastructure could take the form of a database that requires humans to periodically check sensor readings, plus digital twins of each WC and patient to keep their states up-to-date, along with some decision making capabilities to notify technicians, physical therapists, occupational therapists, caregivers, or physicians, as appropriate.

Further, patient needs can be analyzed, and some service scenarios proposed. While these are developed, corresponding WC components should be identified as resources to enable them. A basic service could include periodically notifying the patient's caregiver that the WC should be inspected, where the notification could be a phone call or text message. An alternative is to monitor WC status through sensors on the WC and identify maintenance needs through condition-based monitoring. Then, either the caregiver could be notified or a repair technician could be notified to arrange a visit to fix the WC. Services related to the patient could be similar; sensors on the WC could monitor patient vital signs (e.g., pulse-oxygen sensor) or their operation of the WC. If their ability to operate the

WC seems to be degrading, then messages could be sent to their occupational therapist, physical therapist, or physician, as appropriate.

3 Research Gaps and Proposed Approaches

While there are many potential benefits, there are also many gaps that span multiple disciplines. Imagine a generative design system that provides an interface that allows a user (business, care provider, patient) to define a device by selecting various features. These can be mechanical (materials, load tolerances, etc.), physical features/functions (interfaces, controllers, sensors, etc.) or services (monitoring features, communications, etc.). The system takes the specifications and searches through all of the potential combinations and presents the user with possible customized solutions from which to select that fit the provided criteria.

3.1 Managing Design Evolution and Complexity

Methods for generating variety and customizing products are available in the product family design literature. They need to be extended to address the complexities inherent in CPHS. One specific issue to be addressed is which subsystem would support the most granular customization capabilities. Typically, it is easiest to customize software (i.e., the cyber subsystem), but this may not lead to the most usable products for certain types of customers who may require customized physical subsystems.

An interesting issue arises when we consider a CPHS learning over time, or further, how a CPHS family may evolve over time. Considerations of CPHS variety during the design stage may greatly underpredict the range of behaviors that individual CPH products may learn after long usage by a customer. Designers may foresee some emergent behaviors among CPHS and their users, but not others. The likelihood of missing emergent behaviors in the second or third CPHS generation, for example, falls significantly from the design stage of the first CPHS generation. Significant research is needed to understand the issues surrounding emergent behaviors of CPHS, particularly as they evolve. Such understanding is a prerequisite of research on methods to predict these behaviors and of design to enhance or mitigate them.

A large variety of designs presents other issues. How can the system intelligently sort through thousands upon thousands of potential combinations and know which to present as potential solutions? In particular, the system will need some way to assess the likely usability of a generated design and be able to discard solutions which would not be acceptable.

From a broad perspective, analysis of CPHS and their cyber-physical subsystems is straightforward since these are technical constructs. Cyber systems can be designed and analyzed using the principles of computer science and software engineering. Physical products can be analyzed using engineering principles. However, human subsystems are not easily analyzed before prototypes of products are available.

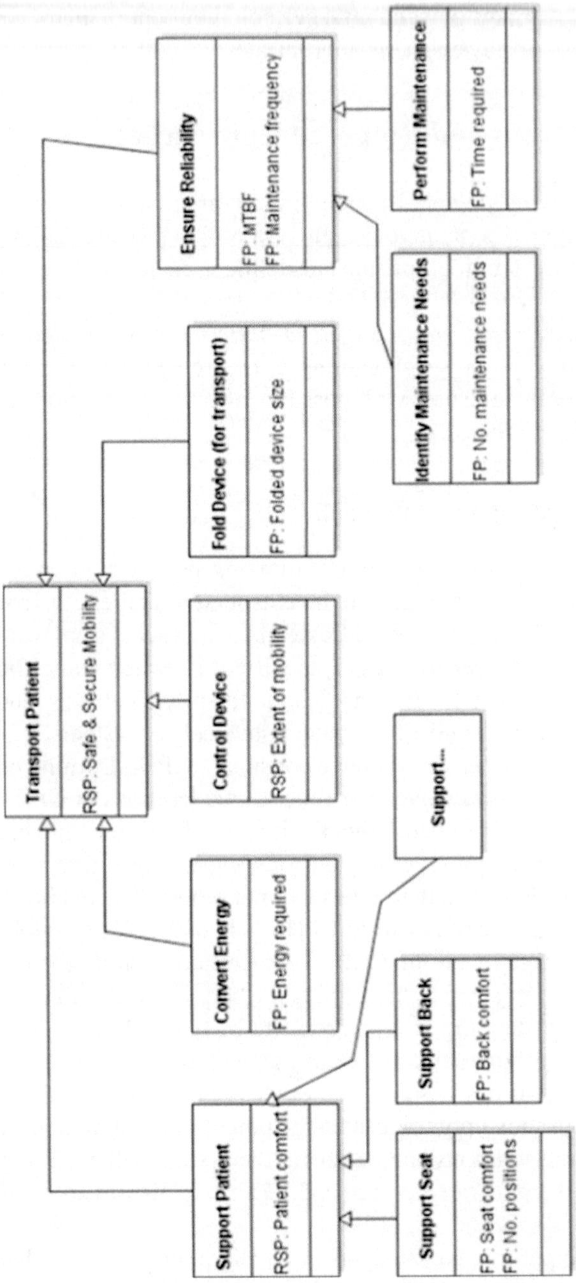

Fig. 2. Function hierarchy for transport patient service, with functional parameters

3.2 Human Voice and Assessing Usability

New knowledge is needed about assessing usability of and human responses to CPHS during their design. This is particularly critical when designing assistive devices. It is not good enough that they simply function. A look into the topic of abandonment will quickly discover examples of devices that seem to be designed 'well' but are rejected at high rates because they are institutional looking, generate stigma because they draw unwanted attention, or that might be desirable and functional but simply make tasks a hassle to perform. Without addressing these types of issues a generative system is likely to perpetuate similar problems with all of the same associated costs to users, insurers and providers.

Decades of research on user-centered design and human-computer interaction have been augmented recently by human-robot interaction research. There has been some progress in the area of automated usability evaluation. However, most of the work in 'automated' testing has been focused on streamlining and automating the collection of data from testers. Generally, usability testing involves recruiting users to use a prototype to perform a task. Information related to the level of usability (effectiveness, efficiency, satisfaction) are collected via a standardized instrument, such as the System Usability Scale (SUS) (Brooke 1996, 2013; Lewis 2018). Most work focuses on automated collection, management and analysis of this data, but still requires engaging users to test each design.

There has been sparse research into methods of programmatically assessing usability using existing data sets, guidelines or heuristics without the direct involvement of users. There is a need to identify methods that are able to quickly assess hundreds or thousands of design possibilities and rank them similarly to how they would be with a human data driven tool such as SUS.

Some work in this direction has been done within very specific contexts, such as usability evaluation of new interfaces for image segmentation systems, such as those used to identify anomalies in medical scans (Amrehn et al. 2019). Another example is the use of a specialized system that uses computer vision to identify elements of a physical interface (a thermostat), infer the function of each element, and apply heuristic rules to assess the likely level of usability (Ponce et al. 2018).

A generalized approach for fully automated usability assessments is yet to be explored. Other fields may contribute to CPHS design, including behavioral psychology and behavioral economics. Both fields seek to understand how humans behave and make decisions under various circumstances. One research direction may be to develop behavioral simulators based on defined personas and existing pools of user collected data that capture shared characteristics of specific user groups. These simulators could be agent-based or based on system dynamics and ideally simulate interactions between the CPHS being designed and each persona that has been defined. Another direction is in the application of machine learning methods. Methods such as Convolutional Neural Networks (LeCun and Bengio 1995) could perform well at identifying usability problems based on heuristics or structured rules. Another direction may be assessments based on Monte-Carlo type simulations to generate distributions of likely behaviors, from which higher level assessments of usability, acceptance, and value could be ascertained.

3.3 Generative Design Alternatives

To illustrate some aspects of generative design, we can consider alternative WC designs that result from the selection of different materials and manufacturing processes. If 3D printing in metal were considered, then complex, optimized frame geometry could be generated, such as that shown in Fig. 3a. On the other hand, 3D printing in polymer could be used to generate different designs. Two alternative concept sketches are shown in Figs. 3b–d, which show a bulky structure (plastics are not as stiff as metal) for the frame in the operating (Fig. 3b) and folded configuration (Fig. 3c), assuming conventional joints among the parts. In contrast, Fig. 3d shows an alternative with compliant (bending) hinges. Large configuration design spaces with various frame, joint, and other WC components may be defined in a straightforward manner.

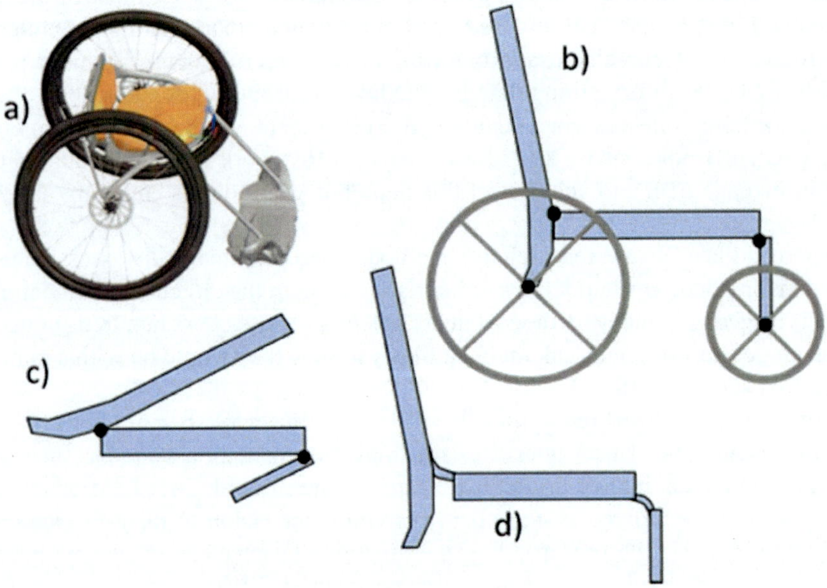

Fig. 3. Wheelchair design concepts, including a) metal 3D printed optimized frame, b) plastic frame with pin joints among parts, c) plastic frame in folded configuration, d) plastic frame with compliant joints among frame segments

Another design consideration is the need to provide customized control designs such as joysticks, tongue-drives, etc. for specific patient conditions to control speed and direction. Other components with different features may be added in order to allow the generation of designs including them. They just need to be added to the system in such a way that all of the inputs, outputs, physical relationships and other required criteria (power requirements, physical connections, available materials, etc.) are defined in relation to other possible components. With a catalog of available components many alternative designs that use one or more of these devices can be developed.

The process of generating the design alternatives is conceptually straightforward. Only the desired functional attributes need to be specified for the product output. The system then can consider all of the possible combinations based on the available library. There may be multiple control schemes that are highly desirable. Each might have different input/output requirements and each option result in different possible combinations of other supporting components. More available components and more requested features can result in exponentially larger combinations of design possibilities (of both components as well as their physical arrangement).

This huge selection of alternatives is one of the chief barriers to such a system. On one hand, a generative system can consider far more possibilities, and very different combinations, than a human designer ever would be able to. Many of these designs could potentially be quite innovative or radical since the relative speed of the system can consider paths that are outside of the realm of more comfortable incremental changes. On the other hand, if the system generates millions of possibilities there is little chance that anyone would see and select one of them as a potential design. This is part of the issue described in Sect. 3.1. The other related issue is emergent behaviors. When many components are integrated, the way that they work together can collectively lead to features and functionality that are not directly provided by any single component but emerge as a result of how they operate in relation to each other. These features can be very (un)desirable. Current generative systems can easily check that components satisfy all inputs, outputs or physical requirements between individual components, but they are not currently able to understand functions that can arise from collective operation.

The other critical barrier is in the area of user acceptance and usability. The same issue of millions of choices is at play here. Ideally, if a potential design is functionally feasible, the next step would be to get an idea of how likely it is to be usable and acceptable. Unlike an automated functional assessment, there are currently no good options for automatically assessing usability. This kind of evaluation should implement checks to ensure that any design conforms to known good practice design heuristics, such as rules for human-computer interaction or proper arrangement of physical controls and displays. This way designs that may violate known human factors considerations can be ruled out.

It is important to note that automated usability assessment as described here would not be a replacement for actual testing. However, some system would be needed to ensure human requirements are considered since it would be impossible to mockup and test all the possibilities generated. An automated usability assessment would serve as an effective filter. A method for assessing the possible usability or acceptability of a generated concept does not yet exist, as compared to a technical assessment.

4 Conclusions

The system described is still an idea, yet many of the prerequisite technologies and systems already exist. Work is ongoing to overcome the barriers. If these issues can be solved, it will be possible to imagine benefits to AT producers (not only of wheelchairs): better provisioning, higher satisfaction and possibly better function. A further integration into so called Industry 4.0 processes can be imagined encompassing many technologies from internet of things, cloud computing, additive manufacturing and other technologies that have an impact on manufacturing. If the components in the generative system are compatible with an automated manufacturing process, it might allow for the specification, ordering and manufacturing of bespoke assistive solutions to be integrated into the provisioning process, potentially bringing innovative solutions that are a better functional and preferential fit for users.

References

Amrehn, M., Steidl, S., Kortekaas, R., Strumia, M., Weingarten, M., et al.: A semi-automated usability evaluation framework for interactive image segmentation systems. Int. J. Biomed. Imaging **2019**, 1464592 (2019)

Autodesk (2022). www.autodesk.com/solutions/generative-design. Accessed 2 Sept 2022

Bendsoe, M.P.: Optimization of Structural Topology, Shape, and Material. Springer, Berlin (1995). https://doi.org/10.1007/978-3-662-03115-5

Brooke, J.: SUS: a "quick and dirty" usability scale. In: Jordan, P.W., et al. (eds.) Usability Evaluation in Industry. Taylor and Francis, London (1996)

Brooke, J.: SUS: a retrospective. J. Usability Stud. **8**(2), 29–40 (2013)

Cook, A.M., Polgar, J.M.: Assistive Technologies: Principles and Practice, 3rd edn. Mosby (2008)

de Domingo, K.: Assistive technologies for cognition: perspectives on funding. Perspect. Neurophysiol. Neurogenic Speech Lang. Disord. **22**(2), 84–89 (2013)

Dicianno, B.E., Joseph, J., Eckstein, S., Zigler, C.K., Quinby, E.J., et al.: The future of the provision process for mobility assistive technology: a survey of providers. Disabil. Rehabil. Assist. Technol. **14**(4), 338–345 (2019)

Dos Santos, A.D.P., Ferrari, A.L.M., Medola, F.O., Sandnes, F.E.: Aesthetics and the perceived stigma of assistive technology for visual impairment. Disabil. Rehabil. Assist. Technol. **17**(2), 152–158 (2022)

Hansen, J.-T., Rosen, D.W.: A product family design method for configuration and spatial layout requirements. J. Comput. Inf. Sci. Eng. **19**, 031006 (2019)

Kairalla, J.A., Winkler, S.L., Feng, H.: Understanding the provision of assistive mobility and daily living devices and service delivery to veterans after stroke. Am. J. Occup. Ther. **70**, 7001290020 (2016)

LeCun, Y., Bengio, Y.: Convolutional networks for images, speech, and time series. In: The Handbook of Brain Theory and Neural Networks, vol. 3361, no. 10, p. 1995 (1995)

Lewis, J.R.: The system usability scale: past, present, and future. Int. J. Hum.-Comput. Interact. **34**(7), 577–590 (2018)

Ponce, P., Balderas, D., Peffer, T., Molina, A.: Deep learning for automatic usability evaluations based on images: a case study of the usability heuristics of thermostats. Energy Build. **163**, 111–120 (2018)

Rosen, D.W., Choi, Y.M.: Extending product family design methods to product-service-system family design. In: International Conference on Engineering Design, Gothenburg, Sweden, 16–20 August 2021 (2021)

Siddique, Z., Rosen, D.W.: On discrete design spaces for the configuration design of product families. Artif. Intell. Eng. Des. Autom. Manuf. **15**, 1–18 (2001)

Sugawara, A.T., Ramos, V.D., Alfieri, F.M., Battistella, L.R.: Abandonment of assistive products: assessing abandonment levels and factors that impact on it. Disabil. Rehabil. Assist. Technol. **13**(7), 716–723 (2018)

Yilma, B.A., Panetto, H., Naudet, Y.: Systemic formalisation of cyber-physical-social system (CPSS): a systematic literature review. Comput. Ind. **129**, 103458 (2021)

Fostering Co-creation Tools for More Inclusive Digital Mobility Services: The DIGNITY Toolkit

N. Carmona[1,2](✉), B. Lazzarini[1], and E. Roca[1]

[1] Institute for Sustainability Science and Technology, Universitat Politècnica de Catalunya, Barcelona, Spain
{boris.lazzarini,elisabet.roca}@upc.edu
[2] Innovation Consultancy Department, Fundació Eurecat, Barcelona, Spain
natalia.carmona@eurecat.org

Abstract. This work describes the application of user-centred design (UCD) and co-creation tools in the design and development process of an online toolkit that aims to tackle the digital divide in the mobility sector. UCD and co-creation approaches aim to address and understand users' needs. This paper details the process of co-creation of the toolkit and reflects on its current results. The design process was based on a first phase of understanding the context and defining the users and their needs, followed by a second phase of co-creating the solution and a final phase of evaluating and testing the toolkit usage and co-creation process. A set of mixed methods employed based on UCD and involving different target groups as end-users has resulted in qualitative data that will feed and strengthen the process of co-creation and the final implementation of the toolkit. Beyond the set of tools presented on the website, the approach followed promotes co-creation, collaboration, and participation as key elements to enhance inclusion and accessibility in mobility ecosystems.

Keywords: Toolkit · User-centred design · Co-creation · Inclusive design

1 Introduction

Digital development is causing a paradigm shift in mobility as in other areas of our daily lives. However, it may exclude those who do not have the digital skills or access to infrastructure, leading to greater social inequalities may and limit sustainable urban development in different dimensions. Some groups – such as people with low levels of education or with low income, elderly people, rural inhabitants, migrants, or disabled people (DIGNITY 2020) – may be vulnerable to exclusion due to their limited access to and use of this technology.

Scientific literature identifies various digital gaps in digitally complex mobility systems, which prove that large parts of the population cannot access or properly operate in the current transport services, lacking the means or the required knowledge and skills (Hoeke et al. 2020). Understanding and promoting how to integrate inclusiveness in digital mobility services is one of the research objectives of the DIGNITY H2020 initiative,

J. Goodman-Deane et al. (Eds.): CWUAAT 2023, *Design for Sustainable Inclusion*, pp. 169–177, 2023.
https://doi.org/10.1007/978-3-031-28528-8_18

a European initiative funded as part of the European Union's Horizon 2020 research and innovation programme.

Co-creation helps to understand and address users' needs by involving target users during the design process. It is a term whose usage and application have increased in many interdisciplinary fields. Co-creation indicates new modes of engagement between people to either create shared value or unleash the creative potential of diverse groups (Rill and Hämäläinen 2018). This definition is aligned with the aim of the DIGNITY project, which considers the diversity of people with a view to an inclusive mobility system.

Several methodologies were analysed and evaluated to contribute to the design of a tangible output: an educational toolkit addressed to different stakeholders, such as policymakers, researchers, and mobility providers, targeting different levels in the mobility ecosystem to meet the needs of potential target users, especially the vulnerable-to-exclusion groups. According to the American Library Association's (n.d.) definition, toolkits are meant to offer practical advice and guidance regarding an issue of concern or importance – especially when the issue is emerging or evolving, and well-established processes for addressing it are not yet widely adopted.

This paper details the design process and implementation of a co-created digital toolkit. Specifically, a combination of methods, such as interviews, benchmarking, and prototyping, among other tools, have been used during the process. These have been essential to identify targeted groups' needs and to co-create the toolkit with several relevant stakeholders.

2 Methodology

This research followed mixed research methods. The overall process included UCD methods to define and understand users' needs in order to design an online toolkit. The phases followed are an integration of the DIGNITY approach, which is based on a three-phase iterative approach. The three phases (framing, bridging, and evaluating) have been maintained since their aim fits this project, but some other tools have been adapted or replaced by others more suitable for this research.

1. Framing: State-of-the-art analysis, which can help to understand the context, gather information, and define the users and their needs.
2. Bridging: Building the solution through co-creating an online toolkit with relevant stakeholders by collecting meaningful insights and feedback.
3. Evaluating: Evaluate the co-creation process, test the toolkit by making sure that it addresses users' needs and helps to tackle the digital gap in the mobility ecosystem (Table 1).

Table 1. DIGNITY toolkit methodology

Phase	Method	Objective
Framing	Literature review	Collect and synthesise research
	Toolkit benchmarking	Identify relevant existing toolkits, highlighting their good practices and pain points
	Stakeholders map	Define stakeholders involved during the co-creation process
	Interview	Understand target users' needs and requirements
Bridging	Co-creation workshop	Obtain feedback from different stakeholders about specific topics
	Web architecture	Definition of the structure of the website
	Interviews	Co-create the proposal, and keep stakeholders updated about the development process
	List of requirements	List of questions about design specifications
	Prototype	Quick prototype of the toolkit website and its content
Evaluating	Accessibility evaluation	List of recommendations based on guidelines and standards
	Review workshop	Workshop to validate the DIGNITY toolkit proposal
	Feedback survey	Adding feedback features to obtain feedback from target users

3 Results on the Process

An initial literature review was done to build a state-of-the-art analysis and find relevant concepts and insights regarding toolkits, inclusive design, and co-creation processes. Garcia-Lopez et al. (2019) presented their design process and implementation of a design knowledge and tools repository (known as Design Toolkit) following a user-centred design approach, which has proven to be successful in the educational field.

After this first step, the benchmarking tool helped to identify several best practices of current toolkits. For this, an initial review was made of 16 existing online toolkits (highlighting their strong and weak points), of which 9 were selected for a deeper comparison and analysis of their main characteristics (topic, targeted users, format, navigation, content, and functions). The toolkits analysed were Design Toolkit (Universitat Oberta de Catalunya n.d.), Design Kit (IDEO 2015), Service Design Tools (Tassi et al. n.d.), UNaLAB Toolkit - Tools for Co-creation (UNaLAB n.d.), Going Digital Toolkit (OECD 2022), Delivering the circular economy: a toolkit for policymakers (Ellen MacArthur Foundation 2015), Inclusive Digital Mobility Toolbox (INDIMO 2022), Inclusive Design Toolkit (University of Cambridge Engineering Design Centre 2017) and Biomimicry Design Toolbox (Biomimicry Institute 2015).

The result of this comparison showed that most toolkits with a wide range of methods and resources tend to have more complex functions and classifications (filter systems, content classified by stakeholders' roles, duration, among others). Most of the toolkits analysed were addressed to specific targeted end-users, such as designers or policy-makers. This specification translates into the need of a specific format and vocabulary adapted to each target group's needs. Furthermore, all toolkits identified have an online website format, but some also provide downloadable versions, which can help to reach more users. This review also collected different functionalities, such as a web searcher, a translation button, user feedback questionnaire or social media share buttons. The most relevant function featured was an accessibility feature that allowed the user to change the text size or colour mode or reset the accessibility options placed in the home page.

Next, a stakeholder's map helped to identify relevant actors and target groups and to list the needs and benefits that each stakeholder could have regarding the DIGNITY toolkit. This mapping supported the selection of stakeholders to be interviewed, since the interviews conducted were a key insight-gathering tool.

The first interview was conducted with a senior designer who has expertise in creating toolkits. He explained how they co-designed their toolkit, its impact, and its limitations throughout the process. Then, a series of interviews were conducted with potential target users previously identified: political and technical policymakers. An interview took place with a former policymaker with expertise in mobility and promoting participatory processes. This helped to collect insights about which content and data the toolkit could supply for her specific needs as a targeted user. Other interviews were arranged with technical mobility policymakers from five mobility organizations, which also helped to understand and define their needs and interests.

These interviews confirmed the need to organise the content in different levels of deepness: a first overview of the tools highlighting the key information, and a second layer of information with all the content available related to the tool implementation process. Case studies were also suggested to be included by both interviewee profiles, due the importance of showcasing previous experiences to implement new strategies and tools.

A co-creation workshop was essential to capture meaningful feedback from several stakeholders. The workshop was carried out in Leuven, Belgium, and participants involved in it had different profiles: mobility operators and providers, policymakers, experts, and researchers. The workshop methodology followed was inspired by the World Café (2022). Participants were split into groups, and each group was assigned to a table. Each table had a facilitator who guided the discussion and collected the feedback. Participants had to discuss four different topics: target users, content, format and design and inclusive design.

Results were usually shared at the end of the session by the facilitators. The first topic helped to define the target users: policymakers, mobility providers, researchers, and vulnerable-to-exclusion' representatives. Some of their needs and requirements were listed, such as involving different impact levels and departments of policymakers or to consider vulnerable-to-exclusion representatives since data and content can help them to put pressure to tackle the digital gap and create a lobby.

The second topic covered aspects related to the content, such as the relevance of showing that the DIGNITY approach was iterative through a visual and interactive wheel figure placed on the home page's website. It was suggested that it was crucial to share relevant data to create urgency to tackle the problem framed. Some content proposals that were similar to those made by the previous interviewees were mentioned: guidelines and templates for each tool implementation, case studies, a literature list, and quotes from target users during the co-creation processes.

The third discussion topic highlighted the need for a format that could be simple and easy to use to ensure the engagement of users, allowing an early overview of what tools can provide.

The fourth topic covered inclusive design. The main toolkit requirement mentioned was to be easy to navigate to the relevant information for a particular user by having a clear understanding of target users' needs. Functionalities and actions to ensure the accessibility of the toolkit were suggested by stakeholders, such as screen readers and contrast ratios tools. Other available resources to evaluate the accessibility were WCAG guidelines (Web Accessibility Initiative 2022) and the Inclusive Design Toolkit (University of Cambridge Engineering Design Centre, n.d.).

All insights provided by the previous tools helped to define a first proposal of the web architecture, which provided the structure of the organised content. The process started by creating a content inventory through a mind map of the content considering feedback previously obtained from stakeholders. Then this information was grouped and labelled, according to the design requirements of the toolkit. After this first iteration, a new version was created in a collaborative online tool to co-create and evaluate this structure with several stakeholders.

Another relevant tool used was a list of requirements (van Boeijen et al. 2014) which consisted of a set of questions based on Pugh's checklist regarding different topics. This list helped to consider and debate with stakeholders some aspects that had not been previously considered throughout the design process topics such as the toolkit's maintenance, costs, safety, testing and standards, among other requirements (Fig. 1).

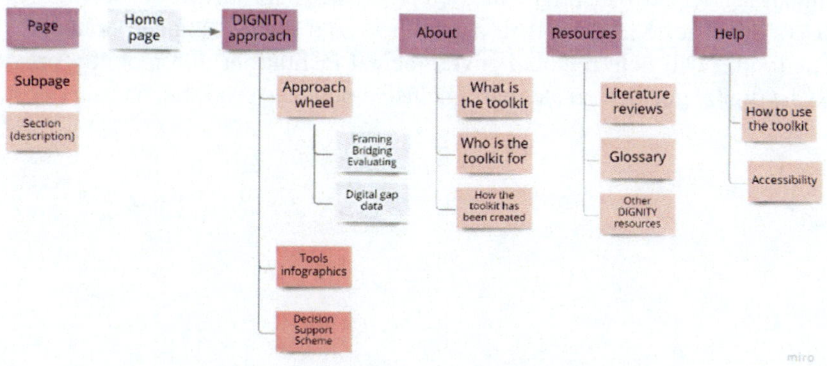

Fig. 1. DIGNITY Toolkit web architecture

3.1 Prototype

The first prototype was based on wireframes that represented an early model of the design. Collaborative online platforms helped to co-create this prototype with stakeholders involved, to build a proposal of how the toolkit should look and which content should it contain, which evolved into an advanced website design.

The DIGNITY Toolkit will be an online platform that contains digital gap assessment tools. The toolkit is targeted on policymakers, mobility providers, researchers, and vulnerable-to-exclusion' representatives, but not excluding other target groups that could be interested in addressing the digital gap.

To start navigating and explore the approach and tools, a visual and iterative wheel is displayed as shown below. The home page also includes relevant data about the digital gap to highlight its urgency. Other website sections included are 'About' (to describe what the toolkit is), 'Tools' (access to the phases and tools to tackle the digital gap) and 'Decision support tool' (test to define which tools are more suitable to and end-user's particular context and needs) (Fig. 2).

3.2 Further Steps

To ensure the success of the DIGNITY Toolkit, it is essential to carry out an evaluation of the design process and implementation. The evaluation phase of this process has not been completed yet since the project is still ongoing and will be carried out during its final phase. The proposal for this phase is to perform an accessibility evaluation of the website. Co-creation workshop participants suggested W3C Web Accessibility Initiative (2022) as guidelines to ensure that the toolkit is accessible. These guidelines help to examine how well the accessibility criteria are met, taking the needs of all the stakeholders and target users into account. It is also important to consider experts and target users to ensure that requirements and users' needs are met.

A review workshop is also expected to display the results of the project, and to discuss outcomes and further exploitation considering feedback from stakeholders. The benchmarking helped to identify it as good practice to introduce a questionnaire to collect end users' feedback and share their experience, so it should be incorporated.

The toolkit will help regional governmental institutions and mobility providers to tackle the digital gap and create more inclusive policies, products, and services.

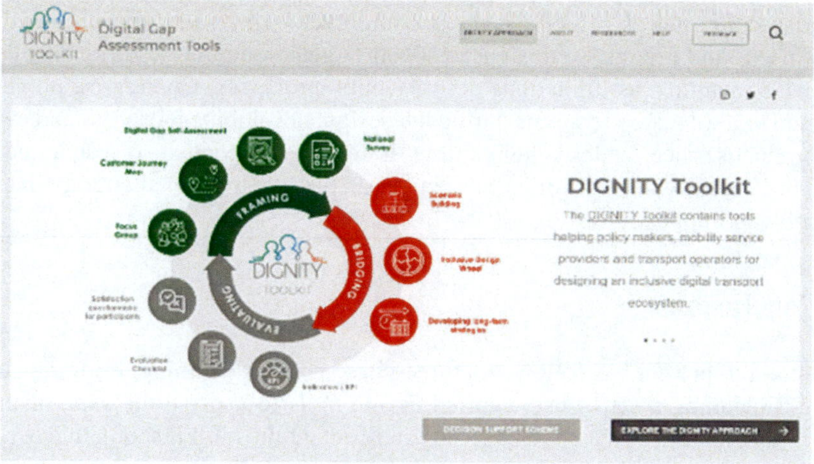

a)

b)

Fig. 2. DIGNITY Toolkit prototypes: a) first prototype, b) latest prototype

4 Discussion

Applying methods and tools based on UCD and co-creation approaches has demonstrated the added value of involving target users throughout the process. Co-creation results in an appropriate model to ensure the inclusion of diversity, focusing on 'vulnerable-to-exclusion' groups during the design process, as it opens up a wide range of voices and experiences that would rarely be involved, and who probably matter the most. Moreover, the societal challenges we are currently facing, such as climate change, establish the need for setting co-creation and participation processes on the political agenda.

Co-creation and UCD tools also help to improve products and services based on end-users' needs and desires, so this ensures well-targeted outputs. Additionally, costs and time were identified as limitations of co-creation processes. Co-creative processes are expected to have many iterations during the design, development process and evaluation of a product/service, which requires time. It depends on context: research and experts might be flexible with these limitations, but it can become more challenging for technical and political implications.

5 Conclusions

This research project has followed a three-phase process (framing, bridging, and evaluating), including a set of user-centred design tools to analyse the state of the art of toolkits and identify what characteristics and content the toolkit should have, considering stakeholders' needs and design requirements. Overall, from a process perspective, co-creation has facilitated the toolkit's design, considering the insights and design proposals validated by stakeholders. So, co-creation processes have been useful and have favoured the inclusion of not just target users, but different stakeholders.

Acknowledgments. This research was done as part of the Dignity project, which received funding from the European Union's Horizon 2020 Research and Innovation Programme under Grant Agreement number 875542.

References

American Library Association: Toolkit creation guide. Yalsa (n.d.). www.ala.org/yalsa/sites/ala.org.yalsa/files/content/ToolkitCreationGuide.pdf. Accessed 18 Jan 2023

Biomimicry Institute: Biomimicry toolbox (2015). https://toolbox.biomimicry.org/. Accessed 18 Jan 2023

DIGNITY: DIGNITY - Digital transport in and for society (2022). www.dignity-project.eu/. Accessed 18 Jan 2023

Ellen MacArthur Foundation: Delivering the circular economy: A toolkit for policymakers (2015). https://ellenmacarthurfoundation.org/a-toolkit-for-policymakers. Accessed 18 Jan 2023

Garcia-Lopez, C., Tesconi, S., Mor, E.: Designing design resources: from contents to tools. In: Kurosu, M. (ed.) HCII 2019. LNCS, vol. 11566, pp. 87–100. Springer, Cham (2019). https://doi.org/10.1007/978-3-030-22646-6_7

Hoeke, L., Noteborn, C., Goncalves, M., Nesterova, N.: DIGNITY D1.1 literature review. Effects of digitalization in mobility in society (2020). www.dignity-project.eu/wp-content/uploads/2020/10/200519-D1.1-Literature-Review-Final.pdf. Accessed 18 Jan 2023

INDIMO: INDIMO inclusive digital mobility toolbox (2022). www.indimoproject.eu/indimo-digital-mobility-toolbox/. Accessed 18 Jan 2023

IDEO: Design kit (2015). www.designkit.org/methods. Accessed 18 Jan 2023

OECD: Going digital toolkit (2022). https://goingdigital.oecd.org/. Accessed 18 Jan 2023

Rill, B.R., Hämäläinen, M.M.: The Art of Co-creation: A Guidebook for Practitioners. Palgrave Macmillan, Cham (2018)

Tassi, R., et al.: Service design tools (n.d.). https://servicedesigntools.org/. Accessed 18 Jan 2023

The World Café World café method (2022). http://theworldcafe.com/key-concepts-resources/world-cafe-method/. Accessed 18 Jan 2023

UNaLAB: UNaLAB toolkit (n.d.). https://unalab.enoll.org/. Accessed 18 Jan 2023

University of Cambridge Engineering Design Centre: Inclusive design toolkit (2017). www.inclusivedesigntoolkit.com/. Accessed 18 Jan 2023

Universitat Oberta de Catalunya: Design toolkit (n.d.). http://design-toolkit.recursos.uoc.edu/es/. Accessed 18 Jan 2023

van Boeijen, A., Daalhuizen, J., van der Schoor, R., Zijlstra, J.: Delft design guide: Design strategies and methods (2014)

Web Accessibility Initiative: Evaluating Web accessibility overview (2022). www.w3.org/WAI/test-evaluate/. Accessed 18 Jan 20

Creating Inclusive Materials and Methods for Co-designing Health Information Technologies with People Who Have Down Syndrome

R. E. Wood[1(✉)], J. Lazar[1], J. H. Feng[2], and A. Forsythe-Korzeniewicz[3]

[1] Maryland Institute for Digital Accessibility, University of Maryland, College Park, USA
{rewood,jlazar}@umd.edu

[2] Department of Computer and Information Sciences, Towson University, and On Sabbatical, Trace Research and Development Center, College of Information Studies (iSchool), University of Maryland, College Park, USA
jfeng@towson.edu

[3] Self-Advocate, Baltimore, USA

Abstract. Medical care disparities for people with Down Syndrome may be exacerbated by the inaccessibility of health systems and personal health data trackers, which can limit their ability to manage their own health. Unfortunately, many of these technologies are created without people with Down Syndrome being factored in as eventual end-users. The few health systems and personal health data devices designed with people with Down Syndrome in mind are more commonly developed *on behalf of* people with Down Syndrome rather than directly involving them in research, design, and development. One reason for less frequent involvement may be a lack of methodological guidance in the literature on how to make participant-facing materials more accessible and research protocols more inclusive to them by playing to their strengths. This paper describes some of the initial efforts of the Health Data Storytelling project. We describe our efforts to make accessible research materials and develop inclusive study procedures to better facilitate the participation of people with Down Syndrome in the design and development of future personal health data and information systems.

1 Introduction

People with intellectual and developmental disabilities, such as Down Syndrome, experience greater health disparities than their typically developing peers (Alshammari et al. 2018). As technology has the power not only to mirror social inequalities, but also to magnify them (Van Deursen and Van Dijk 2014), these disparities may be further exacerbated in patient-facing electronic medical record systems and personal health trackers because of the inaccessible way personal health data and information is presented to people with Down Syndrome.

A. Forsythe-Korzeniewicz---No University Affiliation (Self-Advocate)

J. Goodman-Deane et al. (Eds.): CWUAAT 2023, *Design for Sustainable Inclusion*, pp. 178–187, 2023.
https://doi.org/10.1007/978-3-031-28528-8_19

Health technologies with poor usability and accessibility are often cited as barriers to the adoption of eHealth systems for people with intellectual and developmental disabilities (Alshammari et al. 2018; Schreiweis et al. 2019). These barriers may impede them from actively participating in decision-making and advocating for themselves.

Historically, technology design and development has not included people with IDDs to the same degree as those with perceptual or motor disabilities (Lazar et al. 2017). One reason that people with Down Syndrome or other intellectual and developmental disabilities may be included less frequently in health technology development processes is the lack of clear guidance on how to make research activities accessible to them from the outset.

This paper reports on an ongoing, multi-phase study: the Health Data Storytelling Research project. The Health Data Storytelling study has two goals: 1) to investigate how to make technology design activities more inclusive for people with Down Syndrome and 2) to co-create accessible presentations of health information in patient-facing health platforms to better support people with Down Syndrome as they manage their health. Iter describes the initial steps we have taken during the first of four phases of research, design, and development. In addition to our co-author (A.F.K.), eight main study participants with Down Syndrome used the materials described on which these observations were based. Below we discuss how we have involved representative people during the research study design process and made participant-facing materials more accessible for people with Down Syndrome by leveraging their strengths.

2 Background

Down Syndrome is a commonly occurring genetic disorder that impacts chromosome 21 (GDSF 2021). Current global estimates indicate one baby out of every 1,000 to 1,100 is born with Down Syndrome (UN 2017). It can impact a person's development, cognition, perceptual and sensory processing abilities, as well as their gross and fine motor skills (Feng et al. 2010). Many people with Down Syndrome have mild to moderate intellectual limitations, which can worsen as they age (Carr 1985). People with Down Syndrome often have co-existing conditions that can affect their health and other functional abilities more frequently than the general population (Bull 2020). The quality of care that people with Down Syndrome receive from individual or multiple healthcare providers is critical to their overall quality of life.

People with Down Syndrome have a strong visual awareness. Using signs, gestures, written words, pictorial, concrete and practical support materials may help them to learn better (Faragher et al. 2020). People with Down Syndrome tend to have better receptive language skills (understanding what others are saying) than expressive language skills (verbalising their own thoughts) (Feng et al. 2010). Faragher et al. found that their attention, short-term, and working-memory difficulties may reduce their ability to concentrate, consolidate, retain, generalise, sequence and use abstract reasoning. Despite struggling with numeracy (i.e., performing mental calculations without aids), people with Down Syndrome are still logical (Faragher et al. 2020). Experiences with future health systems need to be engaging and play to their strengths (Hatzigiannakoglou 2015; Lazar et al. 2018).

Existing literature related to the development of systems and devices for people with Down Syndrome often describes how they are developed *on their behalf of* rather than directly involving them in research. Projects like Junk Food Destroyer consulted the literature rather than people with Down Syndrome (Hatzigiannakoglou 2015). Most studies use observations of people with Down Syndrome supplemented with "field notes, questionnaire, surveys, semi-structured interview from primary stakeholders (parents, caregivers, teachers, experts) to *fill in the gaps*" (Shahid et al. 2021). This was the case in Mohammedi and Augusto's (2020) health app, which used User-Centred Intelligent Environments Development Process and Down Syndrome-specific heuristics to create personas followed by interviews with various people with Down Syndrome, parents, and administrators, which they used to create a prototype *for them*. In contrast, in a different, nutritional, app, people with Down Syndrome took part in co-design workshops (Lazar et al. 2018). While participants did not physically design the prototype, they took part in activities that directly informed those designs. The Web Central Fun project used a "very inclusive celebratory approach" during usability testing (Kirijian et al. 2007). That project included recommendations for future systems, such as: begin with fun activities to establish a pattern of success or "avoid conditional phrasing."

One reason for this may be a lack of literature detailing how to adapt technology design activities specifically for people with Down Syndrome. In fact, only one set of guidelines exists: the Inclusive Design for Down Syndrome (ID4DS) Methodology (Macias et al. 2018). Broader accessibility guidelines do exist, such as the Cognitive and Learning Disabilities Accessibility Task Force (COGA TF). However, guidelines, like COGA, cover several highly heterogeneous populations (i.e., cognitive, intellectual, developmental, learning, and language disabilities). While some suggestions may be helpful, others are less applicable to the Down Syndrome population.

Other recommendations tend to be piecemeal. They may apply to specific methods (e.g., surveys by Liljenquist et al. (2019)), or are technology design recommendations with implications for content presentation (e.g., Kirijian et al. 2007). Unfortunately, none of these studies provided sufficient methodological detail or concrete examples for researchers trying to make the specific technology design activities or study documents accessible to people with Down Syndrome.

3 Methods

The Health Data Storytelling research study follows a technology development lifecycle spread across four phases (i.e., semi-structured interviews, large-scale survey, co-design workshops, usability study with eye-tracking). To ensure our research study was accessible to our participants from the beginning, we took a two-prong approach. Prior to its design we conducted informal interviews with three subject matter experts: a medical doctor who runs a Down Syndrome-focused clinic, an occupational therapist who uses AAC board imagery to create health stories as a communication tool with patients (Frank 2017), and a professor whose work focuses on mathematics education of people with Down Syndrome. The interview results refined the research scope to be more relevant to the population and informed the initial accessible document recommendations.

The second prong involved a self-advocate with Down Syndrome in consulting with us and helping us to make our study materials and processes more accessible. He reviewed all participant-facing documents and made suggestions for accessibility improvements and question wording. He also tested all study activities that future study partners with Down Syndrome would also do, to likewise improve the accessibility of research activities. Finally, he provided feedback on how methodological procedures could better align with the strengths of people with Down Syndrome. Following his recommendations and those of the three subject matter experts, the initial research study application was revised, submitted, and later approved by the university's Institutional Review Board.

We used the two-prong approach to progressively develop our accessible content. By taking this approach, the first draft of research materials and procedures was informed not only by the existing (albeit limited) literature of best practices, but also be the practical experience of subject matter experts, who work with people with Down Syndrome on a daily basis. This helped to make the first draft of the content and processes somewhat accessible and inclusive before going through the iterative revisions with our lived-experience expert. This appeared to be a practical and efficient way of quickly developing more accessible and inclusive research materials.

We converted most participant-facing content into full-screen presentations (Fig. 1). For each page or slide, we used simple, plain language to support varying reading abilities. Content was written to build gradually and grouped into logical sections. We read the presented content aloud to support reading comprehension.

Fig. 1. Two consecutive slides visible to potential participants during screening procedures

We used images to support understanding of what was being said aloud by the research team member. Each segment or slide also included at least one support image per sentence. Past work found that people with Down Syndrome have similarly mature stylistic design tastes to their typically developing peers (Feng et al. 2010). Because of this, and given the age of our participants (i.e., 16 + years old), we avoided immature imagery. While images needed to be clear enough to support the text, we wanted to avoid any feelings of infantilisation by image choice. We used a mixture of photographs and illustrations selected from free to use, royalty-free media repositories (e.g., Pexels, Vecteezy).

Assessing and Obtaining Consent and Assent: Determining someone's ability to provide consent occurred while we went through the consent procedures document, but typically prior to a parent (or guardian, if they had one) signing permission to consent on their behalf. The only time assessment of capacity did not occur was when the parent or guardian stated that the individual did not have capacity to consent. Consent and assent documents used simple sentences (i.e., 3rd 4th grade reading level) with support images. All consent and assent documents included the required informed consent information. To our university's IRB, we discussed how minimal content presentation and communication was a matter of providing equitable and accessible content according to the strengths of our population. This gave us flexibility to use reduced wording.

After we had explained each section, we asked section-specific questions to assess capacity to consent. To demonstrate capacity, participant responses to the questions needed to satisfy four conditions: 1) clearly and consistently state it was their *choice* to participate, 2) demonstrate *understanding* 3) show *reasoning* or rational reasons, 4) *appreciate the risks and benefits* from being in the study. If they demonstrated capacity, then the parent (or guardian, if they had one) did not need to fill out a consent form as is best practice with other populations who likewise have conditions that affect their cognition (UC Davis 2002).

As people with Down Syndrome can have difficulty expressing themselves, our consulting self-advocate also suggested that after each section, we asked the individual if the researcher had described the section's content in a way that made sense to them and if they had any questions about what was discussed. This would provide the researcher with an opportunity to re-phrase content or watch for any physical or verbal indicators that the individual with Down Syndrome had questions, or was confused or concerned about something.

Questions were broken up and made relevant to the specific consent section (e.g., procedures, risks, benefits, compensation). As people with Down Syndrome have less short-term and working memory than their typically developing peers (Baddeley and Jarrold 2007), capacity-related questions were asked immediately after we went over the section rather than at the end of the document. For example, after the project description section, we asked: "How would you describe this study to a friend?" Questions were also broken up in this way to minimise the need to recall content from several sections prior. When recall was required, we summarised content from previous sections, such as weighing the risks and benefits of a study to make a considered decision about participation.

Data Analysis: We reflected upon unrecorded study activities (i.e., screening and consent) and an interview with a participant who chose not to be recorded. Next, we reviewed and analysed interview videos across several sessions from the remaining seven participants who consented to being recorded. Research team members then reflected upon how procedures and materials could be improved and identified themes through group discussion until consensus was met.

4 Results

This section describes what we have learned so far in order to create more accessible design research methods and participant-facing documents (i.e., study interest sign-up form, screening protocols, consent forms, demographic survey, interview questions). Below, we describe eight ways to make participant-facing materials and methods more accessible to people with Down Syndrome during research activities.

Methods: Assess and Obtain Consent or Assent. Grouping section content and asking assessment questions immediately after each section was a successful strategy. Scaffolding capacity to consent questions allowed them to build upon their previous responses through gradual reflection of the information as it was presented to them. This format appeared to minimise the reliance upon recall across sections.

Methods: Confirm that Researcher Is Communicating Well. After each section, enquire "Did I explain that clearly?" If something was not clear, rephrase content.

Methods: Encourage Self-advocacy. After the first time a participant advocated for themselves in some way (e.g., asked for a break, or to skip a question), we encouraged self-advocacy by saying things like, "I really like how you advocated for yourself just then. That's exactly what we want you to do whenever you are feeling X." By verbally praising and connecting a participant's boundary setting behaviour with self-advocacy, researchers can foster an environment which empowers participants to be critical of study processes, or of questions being asked. It may even stimulate greater levels of candor during interviews, as we saw in our study. In this way, encouraging self-advocacy can also increase participant expressiveness, which this population can struggle with (Feng et al. 2010).

Methods: Adapt Interview Wording for Relevance and Specificity. Reword responses or questions by clearly linking follow up questions and answers to what the individual said. For example, instead of using the generic term "a tablet," reword the question to mirror their specific response or experience: "your iPad Pro."

Content: Keep Answer Options Minimal, Clear, and Specific. Use large answer buttons so shy or less verbal individuals can point to a response. Minimise the number of response options to 3–4 options. Use a large font as most of our participants wore glasses. This also supported the practice observed during in-person interview sessions, when some participants would use their finger as a guide as they read question aloud to help their comprehension.

Content: Scaffold Question Complexity. Begin a group of related questions with a simple question that required a yes or no response. This can encourage the participant to open up as they begin to recall and progressively reflect upon the topic.

Presentation: Use Images to Support Comprehension and Expression. Using images did support understanding with our participants. Use simple, clear language (e.g., 3rd–4th grade reading level) with age-appropriate images that connect the ideas. Including images in both the question and the answer seemed to support recall via recognition as well as the participant's expressive language abilities.

Presentation: Select Images with Care. Using either photographs or illustrations is fine as long as the image is a clear, literal match. Avoid abstract conceptual links or visual metaphors whenever possible. Instead, imagery should be concrete and obviously support the written language. If possible, have alternative images you can select based on participant's previous response (i.e., dynamically inserting relevant imagery, such as switching out a generic tablet illustration to an iPad Pro).

5 Discussion

Assessing and obtaining consent can be a challenge when conducting research involving people with Down Syndrome, because consent documents and procedures may not be accessible (Shahid et al. 2021). In addition to potentially being exclusionary, the inaccessibility of consent materials and protocols may not meet people with Down Syndromes' actual capacity to consent. Furthermore, it may perpetuate an inaccurate perception within the research community that obtaining consent and demonstrating capacity to consent may be beyond the capabilities of this population.

When first reviewing our study materials and procedures, our consulting self-advocate expressed concerns with the consent process being overly detailed and long. They said it may be too much information for people who might have very different attention, reading, and language skills. They recommended minimising the information as much as possible. Future work should investigate how to reduce informed consent content further.

Reframing confirmation questions during consent (e.g., "Did I explain that section in a way that makes sense?"), as suggested by A.F.K., had several benefits. First, it framed any misunderstanding as a limitation of researcher communication rather than placing the onus of comprehension upon the participant. Second, it let us gauge our participant's unique receptive language abilities more effectively. This interjection provided an opportunity for parents, guardians, or aides to clarify points in the section that may have been confusing. This, in turn, allowed us to then rephrase future content to better align with how the individual communicated.

Even if the individual does not ultimately demonstrate capacity to consent, an added benefit of assessing capacity to consent as described in Sects. 3 and 4 was that it also helped the participant get used to the research team members, the interview format, and advocating for themselves by asking questions as they occurred to them. Researchers may gain insight into how they communicate and can plan future study sessions to be more accessible to that individual's unique abilities and strengths.

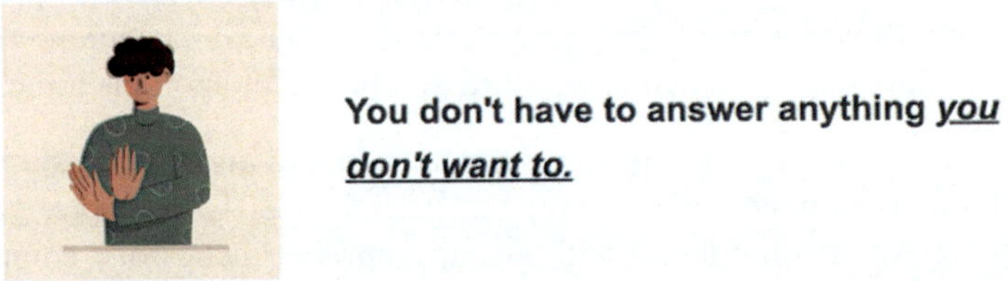

Fig. 2. A person crossing their arms in a "no" movement indicating that answering is voluntary

Support Image Considerations: Selecting appropriate support imagery requires additional testing as interpretation can vary between individuals. In line with past work (Kirijian et al. 2007), no one voiced a preference between photographs and illustrations. However, one of our phase one interview study participants did verbalise confusion during the consent process concerning one image intended to indicate *not having to do* *something* (Fig. 2). In this illustration, the hand position was interpreted as sign language. In recent years, there has been an increased interest in the Down Syndrome community in teaching sign language and finger spelling to people with Down Syndrome to support their expressive communication skills (Powell and Clibbens 1994) as signs and gestures are helpful. Several participants or their parents mentioned using sign language in this way. Researchers should be aware of this practice when selecting appropriate images to supplement text.

Fig. 3. Consecutive questions using scaffolding method to build upon complexity

Scaffold Question Complexity: When asking questions, it is important to slowly build upon complexity. Scaffolding questions allows people with Down Syndrome to link questions with any related follow-ups (Fig. 3). While it is better to minimise the number of response options in follow-up questions, several of our participants did not struggle when a larger number of options (<7) were present. It is, however, important to limit the total number overall to reduce any feelings of being overwhelmed by too many options to process and choose from.

6 Conclusion

To increase the participation of people with Down Syndrome in research, it is critical that it is accessible and uses inclusive practices. Accessible documents and inclusive methods that facilitate communication can better support people with Down Syndrome as participants in directly informing the research, design, and development of health technologies. By consulting with a self-advocate who has Down Syndrome during study planning phases, future researchers can similarly improve the accessibility of their technology research study design and procedures with these suggestions. While it may take additional time and effort during the beginning of a research study to make participant-facing documents more accessible, the steps described were critical to ensure more inclusive methods that supported the strengths of people with Down Syndrome, to actively engage them in research.

Acknowledgements. A special thanks to our subject matter experts: Dr. Brian Skotko M.D., Dr. Rhonda Faragher Ph.D., and Dr. Katie Frank Ph.D., OTR/L. The work reported in this publication was supported, in part, by grant number 90RE5027 (Universal Interface and Information Technology Access RERC) and 90REGE0008 (Inclusive ICT Rehabilitation Engineering Research Center), from the National Institute on Disability, Independent Living, and Rehabilitation Research, U.S. Administration for Community Living, Department of Health and Human Services. Grantees undertaking projects with government sponsorship are encouraged to express freely their findings and conclusions. Points of view or opinions do not, therefore, necessarily represent official policy of the Federal government.

References

Alshammari, M., Doody, O., Richardson, I.: Barriers to the access and use of health information by individuals with intellectual and developmental disability IDD: a review of the literature, pp. 294–298. 2018 ICHI, New York, NY, USA (2018)

Baddeley, A., Jarrold, C.: Working memory and Down Syndrome. J. Intellect. Disabil. Res. **51**(12), 925–931 (2007)

Carr, J.: The development of intelligence, pp. 167–186. In: Lane D, Stratford B Current Approaches to Down's Syndrome. Holt, Rinehart & Winston of Canada Ltd (1985)

Faragher, R., Robertson, P., Bird, G.: International guidelines for the education of learners with Down Syndrome – Down Syndrome International (DSI) (2020). https://workdrive.zohoexternal.com/external/8CkAlci1QBB-Jxfap. Accessed 16 Jan 2023

Frank, K.: People with Down Syndrome – Resources | Adult Down Syndrome Center (2020). https://adscresources.advocatehealth.com/visual-supports/. Accessed 16 Jan 2023

Feng, J., Lazar, J., Kumin, L., Ozok, A.: Computer usage by children with Down Syndrome: challenges and future research. ACM Trans. Access. Comput. (TACCESS) **2**(3), 1–44 (2010)

GDSF: FAQ and facts about Down Syndrome (2021). www.globaldownsyndrome.org/about-down-syndrome/facts-about-down-syndrome/. Accessed 16 Jan 2023

Hatzigiannakoglou, P.: Junk-food destroyer: helping adolescents with Down Syndrome to understand healthy eating through serious game. In: Proceedings of the 2015 VS-Games, Skövde, Sweden, 16–18 September, pp. 1–5 (2015)

Kirijian, A., Myers, M., Charland, S.: Web fun central: Online learning tools for individuals with Down Syndrome, in: Lazar, J. (ed.) Universal Usability: Designing Computer Interfaces for Diverse User Populations, pp. 195–230. Wiley (2007)

Lazar J, Feng JH, Hochheiser, H.: Research Methods in Human-Computer Interaction, 2nd edn., Morgan Kaufmann (2017)

Lazar, J., Woglom, C., Chung, J., Schwartz, A., Hsieh, Y.G., et al.: Co-design process of a smart phone app to help people with Down Syndrome manage their nutritional habits. J. Usability Stud. **13**(2), 73–93 (2018)

Liljenquist, K., Kramer, J., Rossetti, Z., Coster, W.: Content development, accessibility and feasibility of a self-report tool for use in programmes serving youth with cognitive disabilities: the participatory experience survey. Aust. Occup. Ther. J. **66**(4), 490–499 (2019)

Macias, A., Caro, K., Castro, L.A., Domitsu-Kono, M.: ID4IDS methodology: Toward inclusive design for individuals with Down Syndrome, In: Proceedings of the 12th EAI International Conference on Pervasive Computing Technologies for Healthcare, New York, NY, USA, 21–24 May, pp. 263–268 (2018)

Mohammedi, A., Augusto, J.C.: Using technology to encourage a healthier lifestyle in people with Down's syndrome. Univ. Access Inf. Soc. **20**(2), 343–358 (2020). https://doi.org/10.1007/s10209-020-00721-y

Powell, G., Clibbens, J.: Actions speak louder than words: signing and speech intelligibility in adults with Down Syndrome. Down Syndrome Res. Pract. **2**(3), 127–129 (1994)

Schreiweis, B., Pobiruchin, M., Strotbaum, V., Suleder, J., Wiesner, M., et al.: Barriers and facilitators to the implementation of eHealth services: systematic literature analysis. J. Med. Internet Res. **21**(11), e14197 (2019)

Shahid, N.M., Law, E.L.C., Verdezoto, N.: Technology-enhanced support for children with Down Syndrome: A systematic literature review. Int. J. Child-Comput. Interact. **31**, 100340 (2021)

Van Deursen, A.J., Van Dijk, J.A.: The digital divide shifts to differences in usage. New Media Soc. **16**(3), 507–526 (2014)

UN: World down syndrome Day. United Nations. www.un.org/en/observances/down-syndrome-day. Accessed 16 Jan 2023

Education and Training

Effectiveness of the Inclusive Design Wheel in Practice: Lessons from Five Pilot Projects

J. A. Goodman-Deane[1(✉)], M. D. Bradley[1], B. Lazzarini[2], E. Roca[2], S. D. Waller[1], and P. J. Clarkson[1]

[1] Engineering Design Centre, Department of Engineering, University of Cambridge, Cambridge, UK
{jag76,mdb54}@cam.ac.uk, {boris.lazzarini,elisabet.roca}@upc.edu
[2] Institute for Sustainability Science and Technology, Universitat Politècnia de Catalunya, Barcelona, Spain
{boris.lazzarini,elisabet.roca}@upc.edu

Abstract. The Inclusive Design Wheel (IDW) is an established inclusive concept design process, emphasizing iteration and early-stage evaluation. However, a structured evaluation of its effectiveness in practice has not been previously conducted. This paper describes how the IDW process was adapted for use in the design of digital mobility services. It was then trialled on five pilot projects in Italy, Spain, Belgium and the Netherlands. Questionnaires were completed by participants (both pilot partners and end-users) and interviews were held with pilot partners at the end of the project. Initial insights from the pilot experiences and the evaluation are described. The pilot teams found the IDW helpful and particularly valued its help in structuring the design process and offering clear steps and tools. Insights for improving the IDW focus on the importance and role of co-creation, issues with recording iteration and the balance between structure and flexibility. Implications are highlighted for improving the Inclusive Design Wheel and for inclusive design processes more generally.

1 Introduction

There is a growing recognition of the diversity in the population and the consequent importance of inclusive design, i.e., "the design of mainstream products and/or services that are accessible to, and usable by, as many people as reasonably possible" (BSI 2005). However, designing inclusively can be difficult in practice. Effective and practical inclusive design processes and tools are needed to help designers and other stakeholders to put inclusive design into practice.

The Inclusive Design Wheel (IDW) is one such process (Cambridge Engineering Design Centre n.d.). It was developed based on over ten years of research and knowledge transfer to help designers to structure the inclusive concept design process within product design (Clarkson et al. 2007; Waller et al. 2015). The IDW is compatible with other design models such as the Waterfall, Double Diamond, V and Spiral models (Clarkson and Eckert 2005; Design Council 2007). However, it is more specifically focused on concept design and emphasizes iteration and early-stage evaluation (Waller et al. 2015). It also

J. Goodman-Deane et al. (Eds.): CWUAAT 2023, *Design for Sustainable Inclusion*, pp. 191–200, 2023.
https://doi.org/10.1007/978-3-031-28528-8_20

directly includes specific activities that are important for designing more inclusively. The IDW is also compatible with co-creation methods (e.g., Cruickshank et al. 2013), which can be used within the different phases and activities of the IDW.

The IDW is composed of four main phases as shown in Fig. 1 (Waller et al. 2015; Cambridge Engineering Design Centre n.d.). Successive iterations of these phases are used to generate:

1. a clearer understanding of the needs that the product or service should meet;
2. better solutions to meet those needs;
3. stronger evidence that the needs are met.

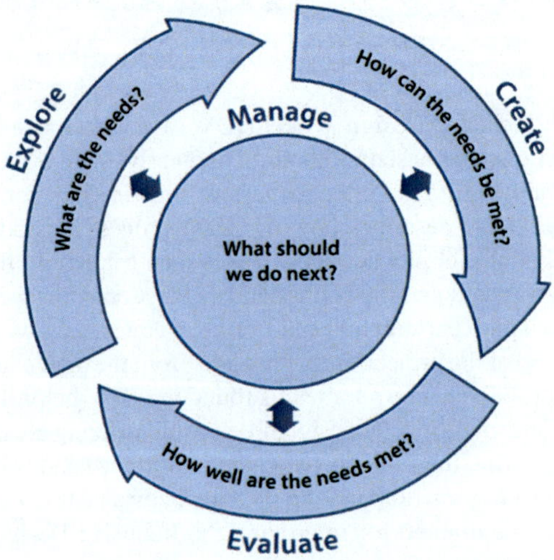

Fig. 1. Overview of the IDW process (from Cambridge Engineering Design Centre n.d.)

The IDW provides a selection of activities within each phase, such as *Create stakeholder map, Observe users, Develop concepts* and *Test with users*. A more detailed diagram listing all the activities can be found in (Waller et al. 2015; Cambridge Engineering Design Centre n.d.).

The IDW process and associated tools have been used on projects from a wide variety of domains and industry sectors (e.g., Centre for Business Innovation n.d.), but a structured evaluation of its effectiveness in practice has not previously been conducted. This paper reports on the use and evaluation of an adapted version of the IDW in five pilot projects across Europe. Section 2 explains how the IDW was adapted for use in the design of digital mobility services and Sect. 3 briefly describes the pilot projects. Section 4 describes how the work was evaluated and presents initial results and insights from the evaluation.

2 Adaptation of the Inclusive Design Wheel

The IDW process was originally intended for use in product design, but it can be adapted for use in different sectors, and for different types of applications. The overall process (shown in Fig. 1) stays the same, but individual activities can be adapted and additional activities added to better match the characteristics and needs of the sector. The process can also be adapted for use in a particular project or company to fit better with an established design process or activities.

As part of the EU funded DIGital traNsport In and for socieTY (DIGNITY) project (DIGNITY project n.d.), the IDW was adapted for use in the design of digital mobility services (Bradley and Goodman-Deane 2021), i.e., transport services that use digital interfaces to provide information or functionality to the user. Examples include websites and apps that provide information about travel, online ticket booking services and digital interfaces for ride sharing and vehicle hire. The digital mobility sector differs from traditional product design in various ways. For example, digital mobility services operate within larger transport ecosystems that involve many different services, providers and commercial, governmental and public stakeholders. Funding and management for public transport services may also differ from those in traditional product design.

Adaptations were also made to the IDW process to fit better with the other activities in the DIGNITY project and to incorporate co-creation methods within the *Create* phase.

A diagram of the adapted IDW process is shown in Fig. 2. Examples of the changes in the IDW include replacing *Observe users* and *Describe user journeys* with *Examine customer journeys*. This was done to tie in with other work on the DIGNITY project that uses customer journey mapping to understand users' experiences and needs (van Apeldoorn and Herpen 2022). Another example is the addition of *Make storyboards* to the *Create* phase. This was included because storyboards are often more suitable than prototypes for describing concepts for a mobility service.

A key issue in adapting the IDW was deciding what to include as an explicit activity in the process diagram (the small circles in Fig. 2). There was only room for a limited number of such activities. As a result, some activities were included as part of a more general activity. For example, co-creation workshops are a key part of *Stimulate ideas* but are not directly visible in the process diagram. Care was also needed in trying to describe each activity in a few words so that people viewing the diagram could get a good impression of what was involved.

When advocating the use of a design process, it is helpful if that process is accompanied by a tool for guiding people through the process and capturing the outputs from activities and the rationale for design decisions. The tool can take a variety of forms, such as a website, a bespoke piece of software or a file in standard software, e.g., Microsoft Office. For the adapted IDW, a design log template (see Fig. 3) was developed in PowerPoint. PowerPoint was chosen because most professionals already know how to use it and it allows users to easily add combinations of text and images. It also aids in the presentation of information to others throughout the design process.

Fig. 2. A diagram showing the adapted version of the IDW for the design of digital transport services and for the DIGNITY project. This diagram shows the individual activities within each design phase. Note that it is not necessary to do all of the activities within each of the phases.

The design log included slides for each of the activities, with brief guidance on the activity and templates to fill in with notes and findings, e.g., different types of stakeholders or ideas generated from a co-creation workshop.

3 Pilot Projects

The adapted IDW was piloted in five pilot projects (referred to as 'pilots' in this paper) across four cities/regions in Europe: one each in Ancona (Italy), Barcelona (Spain) and Flanders (Belgium), and two in Tilburg (Netherlands). All of the pilots involved the local authority for the city or region plus other partners such as local transport providers and consultancies with expertise in transport and mobility. More information on the pilots can be found in Goodman-Deane and Bradley (2022).

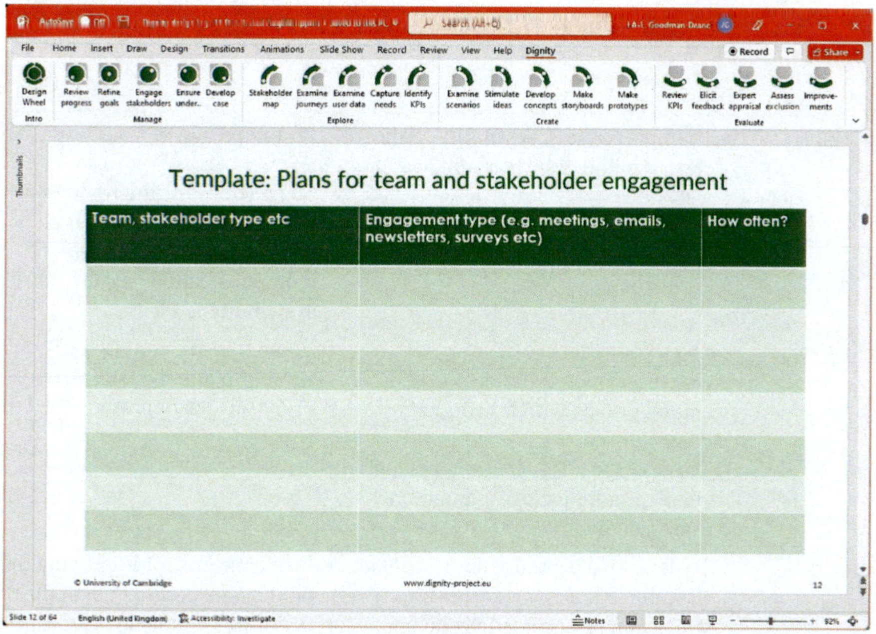

Fig. 3. Example template from the Inclusive Design Log (v1.0) used in the DIGNITY project. The custom navigation bar at the top of the screen was developed for this project.

The pilots used the adapted IDW to develop concepts, prototypes and recommendations for more inclusive mobility services in their regions. Different pilots had different specific aims and developed different outputs, as detailed in Table 1.

This pilot work was conducted by the pilot teams with guidance and an initial evaluation of their concepts provided by the research team from the University of Cambridge. The work started in summer 2021 and this paper reports on work conducted up to May 2022. During this period, all of the pilots did substantial work on the *Manage, Explore* and *Create* phases of the IDW. Four of the five pilots did some *Evaluate* activities as well, thus completing at least one iteration of the adapted IDW process. However, as with any iterative process, the teams are continuing work to further improve their outputs.

The implementation of the adapted IDW varied between pilots, depending on factors such as the scale and type of the problem being addressed and whether the team was adapting an existing solution or building a new one. There were differences in the exact activities conducted, in their implementation and in the type and level of detail of the outputs.

Table 1. Summary of the five pilots

Pilot	Aims	Outputs
Ancona	• To improve the accessibility of the current public transport app and website • To add non-digital solutions to include users who face digital exclusion	• Prototype of a new version of the app • Suggestions for improving the accessibility of the website • Seven (mostly non-digital) recommendations for the transport system in the region
Barcelona	• To promote the use of demand-responsive transport (DRT) among groups with low digital competence • To make the DRT smartphone app simpler, more inclusive and more user-friendly	• Three (mostly non-digital) concepts for improving the DRT service • Proposals for improving the DRT app
Flanders	• To develop a central point of contact (Hoppincentrale) for end users, which can be accessed via an app, website or call centre. This will provide public transport information and help end users to plan trips	• Five concepts for improving the public transport system in the region • A prototype of an app for accessing Hoppincentrale • Proposal for a training programme to equip public transport personnel in assisting people with disabilities
Tilburg pilot 1	• To develop an intervention to help digitally excluded people (particularly those who are older) to get from A to B	• Concept for a central travel information phone number • Five storyboards describing how people could to find out about and get access to this phone line
Tilburg pilot 2	• To develop a concept for a socially and digitally inclusive bike-sharing scheme, with special attention to the needs of migrant women	• Guidelines for bike sharing schemes • High level description of a concept for a new bike sharing service • Recommendations for an equitable shared cycling system in the region

Nevertheless, a key part of all the pilots was the co-creation workshops (see Fig. 4), held as part of the *Stimulate ideas* activity. All pilots held at least one such workshop with target users for their project. The participants usually included some from user groups at increased risk of digital mobility exclusion (such as older people or those with disabilities). All the workshops included elements that identified and explored users' transport needs, as well as ones that elicited ideas for meeting those needs. Most of the workshops were held in person, but one pilot had to run theirs online due to COVID-19 restrictions. To balance this, they also held some telephone interviews with users at higher risk of digital exclusion.

Fig. 4. Photos from some of the co-creation workshops

4 Evaluation

The pilot work on the adapted IDW was evaluated using a range of methods:

- a questionnaire addressed to participants (end-users) in the co-creation workshops (part of the *Stimulate Ideas* activity), conducted at the end of or soon after the workshops;
- a questionnaire addressed to pilot partners, conducted partway through the IDW process (usually somewhere in the *Create* phase);
- a questionnaire addressed to pilot partners, conducted at the end of the IDW work examined in this paper;
- a semi-structured interview with pilot partners, also conducted at the end of the IDW work examined in this paper.

In addition, throughout the process, the University of Cambridge team recorded feedback from the pilots, any issues that arose and their own reflections on the experiences during the pilots.

The questionnaire and interview responses indicated that the pilot partners considered the adapted IDW process to be helpful overall. They particularly valued the help it provided in structuring the design process and offering clear steps and tools. Nevertheless, the experience on the pilots highlighted a range of issues and lessons to learn for improving the IDW. The analysis of the evaluation results is still ongoing but some of the initial lessons are described below.

4.1 Co-creation Workshops

All the pilots found the co-creation workshops helpful. For example, one of the pilot teams said, *'[It was helpful] in giving insights directly from the target group about the target group.'* The participants in the workshops also valued them – the great majority of them found them useful, enjoyed them and felt they generated useful ideas. Furthermore, the observation of the pilots indicated that the co-creation workshops often provided key ideas and insights, both for the *Explore* (understanding users' needs) and the *Create* (coming up with ideas for solutions) phases.

This indicates that these workshops are extremely important for the IDW process. However, they do not appear directly in the process diagram in Fig. 2 because they are part of the *Stimulate ideas* activity. The name of this activity does not convey anything about co-creation or user involvement and, as a result, the co-creation workshop could be easily overlooked. It would help to emphasize the importance of this activity if it appeared directly in the process diagram.

Furthermore, the workshops provided important information for both the *Explore* and *Create* phases. However, they are only located in the *Create* phase in the process diagram. Further work is needed to consider how the relevance of the workshops to both phases can be communicated.

4.2 Iteration

Iteration is an inherent aspect of design in general (Wynn and Clarkson 2018) and of the IDW in particular (see Sect. 1). As well as the macro iteration of the *Explore*, *Create* and *Evaluate* phases, teams often engaged in micro iteration cycles where they made rapid, small changes to their insights and concepts as they discussed and tried out things together.

However, some of the teams found it difficult to record this iteration within the PowerPoint design log: *'[The design log] could be improved by organising better the addition of new inputs ... to record the evolution of the pilot as it is [a] continuous process'*. Some of the pilot teams came up with their own method of tracking iteration within the log, by annotating new material with a coloured circle with the word 'NEW' and the date. Some teams kept both the old and new versions of slides within the design log. However, keeping old versions makes the log more difficult to navigate, and it was still hard to capture and communicate how the content varied over time.

Capturing and communicating successive and iterative updates to multiple interlinked components is a difficult challenge that occurs in other domains as well. A rigorous and structured approach to version numbering can help, but excessive rigour and structure may be onerous and could stifle creativity. Bespoke project and file management software might be able to help, but this requires teams to adopt and learn new software, thus creating an extra barrier to its use. An alternative might be to create custom buttons within standard software (such as PowerPoint) to streamline the process of tagging material with the date. Further work with design teams would be required to explore and refine solutions to this issue.

4.3 Flexibility

Some teams felt that the IDW process was a bit constraining. In the final evaluation questionnaire, one of the teams said, *'[it] seems like you have to complete all the activities in order which is counter intuitive'*. Another commented, *'we used some elements but not the whole IDW method.'*

The adapted IDW was intended to be structured and yet flexible, with multiple iterations, each containing a subset of the activities in Fig. 2. However, this was not adequately conveyed in the process diagram. Furthermore, the use of PowerPoint for the design log meant that the slides for the different activities were presented in a linear order. Alternative tools, such as Adobe Illustrator and Miro, offer some potential to break away from this linear presentation. They allow boards to be laid out and moved around spatially within a workspace. This might fit better with the intended paradigm for completing the activities in the IDW. However, the availability of and prior experience with these tools is lower than PowerPoint, so further work is needed to determine the best environment in which to implement the design log and/or whether PowerPoint can be better customised to suit a more flexible process.

5 Conclusions and Further Work

This paper has reported on initial results from the evaluation of five pilot projects that used an adapted version of the Inclusive Design Wheel in the design of mobility services with a digital element. The pilot teams found the adapted IDW process helpful and particularly valued its help in structuring the design process and offering clear steps and tools.

Particular issues identified in the evaluation include the importance of the co-creation workshops and the need to highlight this better in the process diagram. The importance of these workshops also underlines the importance of early-stage user involvement and co-creation in inclusive design processes more generally. The evaluation also identified issues with recording iteration, which has relevance for both the IDW and for design processes more generally. There is also a tension between providing enough structure that the process is actually helpful and allowing enough flexibility that it can be used in creative ways in a variety of situations.

Adaptations to the IDW process and design log in response to this evaluation are currently underway (at the time of writing in December 2022). The latest versions of both can be found on the Inclusive Design Toolkit at: www.inclusivedesigntoolkit.com/transport_overview.

Acknowledgements. This research was done as part of the DIGNITY project which received funding from the European Union's Horizon 2020 research and innovation programme under grant agreement number 875542. For the purpose of open access, the author has applied a Creative Commons Attribution (CC BY) licence to any Author Accepted Manuscript version arising from this submission.

References

Bradley, M., Goodman-Deane, J.: D2.2 Guidelines for inclusive design processes for digital products and services. DIGNITY project deliverable (2021). www.dignity-project.eu/wp-content/uploads/2022/06/DIGNITY-Deliverable-2.2.pdf. Accessed Sep 2022

BSI: BS 7000–6: 2005 - Design Management Systems. Managing Inclusive Design. Guide – BSI British Standards, The British Standards Institution (2005)

Cambridge Engineering Design Centre: Inclusive design toolkit: Concept design process: Overview. Engineering Design Centre, University of Cambridge, UK (n.d.). www.inclusivedesigntoolkit.com/GS_overview/overview.html. Accessed Sep 2022

Centre for Business Innovation: European Inclusive Design Consortium (n.d.). www.cfbi.com/inclusivedesign.htm. Accessed Sep 2022

Clarkson, J., Coleman, R., Hosking, I., Waller, S. (eds.): Inclusive Design Toolkit, 1st edn. Engineering Design Centre, University of Cambridge, Cambridge (2007)

Clarkson, J., Eckert, C.: Design Process Improvement: A Review of Current Practice. Springer, London (2005)

Cruickshank, L., Coupe, G., Hennesy, D.: Co-Design: fundamental issues and guidelines for designers: beyond the castle case study. Swed. Des. Res. J. (2), 48–57 (2013)

Design Council: 11 lessons: Managing design in 11 global companies (2007) www.designcouncil.org.uk/fileadmin/uploads/dc/Documents/ElevenLessons_Design_Council%2520%25282%2529.pdf. Accessed Sep 2022

DIGNITY project (n.d.) www.dignity-project.eu/. Accessed Sep 2022

Goodman-Deane, J., Bradley, M.: D3.3 report on the meso level process (Inclusive design Wheel). DIGNITY project deliverable (2022). www.dignity-project.eu/wp-content/uploads/2022/07/Dignity-Deliverable-3.3-final.pdf. Accessed Sep 2022

van Apeldoorn, N., Herpen, J.: Customer journey mapping: How vulnerable-to-exclusion groups are coping with the digitalization of mobility. Transp. Res. Procedia 2022 (2022, to appear)

Waller, S., Bradley, M., Hosking, I., Clarkson, P.J.: Making the case for inclusive design. Appl. Ergon. Special Issue Inclus. Design **46**, Part B (January), 297–303 (2015)

Wynn, D.C., Clarkson, P.J.: Process models in design and development. Res. Eng. Design **29**(2), 161–202 (2017). https://doi.org/10.1007/s00163-017-0262-7

Towards a More Inclusive Learning Experience for Minority Ethnic Students in Brunel Design School

H. Dong(✉), Q. Ni, and A. Manohar

College of Engineering, Design and Physical Sciences, Brunel Design School, Brunel University London, London, UK
{Hua.Dong,Qing.Ni,Arthi.Manohar}@brunel.ac.uk

Abstract. Degree award gaps have been reported for every university in the UK, and they are often unexplained. Reducing the award gaps between white and minority ethnic students in the UK is an effort to achieve equality, diversity and inclusion (EDI), an increasingly important agenda for higher education institutions. This paper reports a study at Brunel Design School, focusing on the award gaps and the learning experience of the minority ethnic students. Focus group and individual interviews were conducted with 25 student volunteer participants in Spring 2022, in addition to student EDI group discussions. It was found that English language skills play an important role in their learning experience and assessment outcomes. Despite the lack of training in English academic writing, the majority of the participants preferred reports rather than video/audio assessment. The students' suggestions and the relevant strategies proposed by published studies will be used to address degree gaps, and improve minority ethnic students' learning experience at Brunel Design School, as part of the action research.

1 Equality, Diversity and Inclusion (EDI) and Award Gaps

In 2005, the same year when the British Standard on Managing Inclusive Design (BSI 2005) was published, the Athena Scientific Women's Academic Network (SWAN) charter (Advance HE 2020) marked the first initiative leading to the later Equality, Diversity and Inclusion (EDI) agenda. Nowadays EDI has become ever-important in universities (Wolbring and Lillywhite 2021), developing and implementing EDI policies has become a significant challenge for higher education institutions nationally and internationally (Scott 2020). Similar terms include Equity, Inclusion, Diversity and Access (EIDA), Diversity and Inclusion (D&I), Diversity, Equity and Inclusion (DEI), and Inclusion, Diversity, Equity and Access (IDEA) (Loehwing and Craig 2021).

EDI has become a commitment of the Advance HE Strategy 2021–2024 (AdvanceHE 2022). Acknowledging the importance of diversity should transform how universities operate, not only about who to recruit, but also what to teach and research, and how to support students and employees to achieve their potential (Scott 2020). Minority ethnic students are an important focus of EDI in higher education.

In the United Kingdom (UK), 81.4% of white students and 68.0% of Black, Asian, and other minority ethnic students were awarded good undergraduate degrees (i.e., first or

J. Goodman-Deane et al. (Eds.): CWUAAT 2023, *Design for Sustainable Inclusion*, pp. 201–209, 2023.
https://doi.org/10.1007/978-3-031-28528-8_21

upper second) in 2018/19. The existence of an award gap has been known about for over 25 years, and such gaps have been reported for every university in the UK (Cramer 2021). Brunel University London set the goal of halving the good degree award percentage differences between white and minority ethnic students by 2024/25. In order to achieve this, ten projects were funded in 2021/22. This paper reports one of the projects conducted at Brunel Design School, a school with an international reputation for inclusive design and design for sustainability research, and Made in Brunel (madeinbrunel.com) collaborative teaching excellence. The project aimed at creating a more inclusive learning experience for minority ethnic students, so as to benefit its increasingly diverse learner population sustainably in the long run.

2 Methods

Action research (Mertler 2017) was adopted as the main research method. Learning analytics such as graduate data (Minocha 2021) and interviews were used to understand the issues and identify actions and potential solutions. In addition, student EDI group meetings and discussions were conducted to gain further insights.

In the process, students were involved as participants, designers and co-researchers (Minocha 2021), e.g., the interview questions were co-designed with students' input, PhD students were employed as assistants to help recruit participants (to minimise the potential impact of the teacher-student power relationship), collect and analyse data. The research received ethics approval from the College of Engineering, Design and Physical Sciences' research ethics committee.

The participants were recruited through poster advertisement and at social events at Brunel Design School. Their participation was voluntary, and a five-pound voucher was given to each afterwards as a 'thank you' gift. The sampling did not necessarily represent the ethnic mix of the school.

Typical interview questions include:

- What type of difficulties do you encounter in your course?
- When you encounter difficulties in a course, do you communicate them to the lecturer or why?
- Do you use the University's English study support? How effective is it for your study?
- For teamwork, what challenges do you have in co-working with Caucasian/English speaking students? And with students from your own country/similar culture?
- What do you think about the existing assessment methods?
- How do you feel about audio/video submissions in comparison with written reports?
- What do you think about the learning and teaching environment?
- Are you aware of the degree awarding gaps between white and non-white students? If yes, what do you think might have contributed to these gaps?
- What are the challenges for you to achieve a first or 2:1 degree (if UG students)?/What are the challenges for you to achieve a distinction or merit degree (if PG students)?
- What are your suggestions to address the degree award gaps?

The interviews were conducted in focus groups (2–5 students) or with individuals (depending on the participant's availability and preference) between March and May 2022. They each lasted 45–120 min, and audio was recorded. The recordings were subsequently transcribed; thematic analysis and coding were applied by one researcher, and cross checked by another. Numbers of identified issues were counted.

3 Results

Learning analytics (in this case the degree award data of design graduates of different programmes) suggest that in 2020/21 (studying under the pandemic, with no-detriment policy/'safety net') the ethnicity degree award gap was 35% for BA Industrial Design & Technology students, 12% for BSc Product Design Engineering students, and 37% for BSc Product Design students.

In total 25 student volunteers participated in the interviews (5 conducted as individual interviews), including 16 Chinese students, 4 Indian students, 4 black students, and one Cambodian student. Five were undergraduate students (UG), and twenty were postgraduate students (PG).

The main results are summarised below ('n' represents the number of mentions):

- **Difficulties encountered in the course:**
  - Difficulty understanding n = 16 (including 4 specifying 'lecturer speaking too fast');
  - lack of clarity in course requirement n = 5;
  - online lack of engagement n = 4;
  - lack of English writing training n = 4;
  - culture difference n = 3;
  - short length of the PG course (i.e., only one year) n = 1;
  - group work n = 1.
- **Coping strategies when encountering difficulties:**
  - email or ask lecturers n = 11 (with two mentioning slow/no response);
  - confused but do not know what to do n = 9;
  - figure out by self n = 1;
  - never asked a question outside class n = 1.
- **English language support:**
  - used the support but found it not very effective n = 11;
  - not heard about it/not used it at all n = 9;
  - scared of speaking in language support groups n = 7 (with one specifying the difficulties of communicating with students from different countries);
  - used it and found it useful n = 4.

- **Team work:**
  - language barriers preventing effective team work n = 16;
  - prefer to be mixed with students from different cultures n = 12;
  - tend to rely on an 'interpreter' whose English is better n = 9;
  - online group work extremely difficult n = 1.
- **Current assessment:**
  - assessment criteria not clear n = 9;
  - being unfair/biased n = 8;
  - fair/inclusive n = 5;
  - negatively influenced because of poor English language skills n = 5;
  - lack of connection between theory and practice n = 4.
- **Audio/video submissions in comparison with written reports:**
  - prefer writing/reports n = 15 (with one specifying it would be a 'pain' to record a video);
  - prefer alternative assessment (e.g., audio/video submission) n = 8;
  - do not have a preference n = 2 (with one specifying all 'would be harder for international students').
- **Learning and teaching environment:**
  - supportive staff and learning environment n = 20;
  - (PG students) would like to get training and access to design workshops n = 5;
  - environment lacking 'design feeling' n = 4;
  - need training on academic writing n = 1.
- **Awareness of degree award gaps:**
  - no awareness at all n = 11 (with three guessing language skills might have contributed to the award gaps if they exist);
  - some awareness of the gaps n = 6;
  - do not care much as practical skills are more important for designers n = 3.
- **Challenges to achieve better degree awards:**
  - supervisor allocation (mentioned only by Post Graduate students) n = 7;
  - unclear requirements and assessment criteria n = 4;
  - do not care much n = 3;
  - lack of report writing skills n = 1.

- **Suggestions:**
  - clarification of assignments and assessment criteria n = 3;
  - improve self n = 3;
  - staff being objective and unbiased n = 2;
  - staff offering effective feedback n = 1;
  - staff communicating more with non-native speakers n = 1;
  - enhancing English Language Centre/Programme n = 1;
  - (staff) being patient n = 1;
  - anonymous submission n = 1;
  - offering bridging courses on writing n = 1.

From the two EDI group meetings organised by students (mainly UG), some common issues were identified:

1. students felt there is an obvious divide between students of colour (SoC) and white students. More inclusive/collaborative module teaching can help address this gap.
2. it has been observed by students that staff at times are not patient or empathetic enough towards SoC (due to language issues), e.g. SoCs are treated differently to other students when they miss classes.

Suggestions were proposed by the student EDI groups, e.g.,

- making submissions and markings anonymous;
- SoC helping in modules might help reluctant students engage better in modules;
- students felt pairing staff of colour with SoC for Major Project or Personal Tutors could work well as they connect better;
- making design briefs simpler and more inclusive;
- more diversity in lecture materials rather than focusing on UK-centric contexts;
- guest lectures connecting academic content to industry helps students to see how things are applied in real world.

4 Discussion

It is very difficult to pin down what exactly causes the 'large, unexplained award gap' (Cramer 2021), similar to the fact that 'no one single element of the student experience can be controlled to enhance satisfaction' (Partington 2019). Despite this, the study still gives us good insights into minority ethnic students' learning experience. To our surprise, most of the students were not aware of the degree award gaps, to quote one student: "*I wasn't aware of any awards actually, let alone any award gaps*".

Command of English language seems to play a role in the academic performance and general experience of minority ethnic students: more than half of the participants (n = 16) acknowledged difficulties in understanding lectures; this was mainly because of the language barrier; the lack of English language skills prevents some students' effective participation in team work (e.g., n = 9 relied on an 'interpreter' within the team to help with communication). In addition, assessment outcomes could be negatively influenced because of poor English language skills (n = 5). Some participants (n = 3) perceived English language skills to have contributed to degree award gaps. Although three students said they did not care about degree gaps (because they believed practical skills were more important for designers), another student made the following point:

> "*I don't think 99% of the companies care about their (candidates') degree. Yeah, they don't care about. They care about you. Like, what can you do? And if you can't speak or if your communication skills are not on point, then it's going to be super, super hard to get a job. So, many people in my course, they're not very confident speaking English. And I think if you're in UK here, the chances you're getting hired is so hard.*"

'Confidence' plays a role in communication skills, as well as culture adaptation. For example, one international student whose native language is English but who is "completely new to the country", said,

> "*they (international students whose mother tongue was not English) were reluctant to speak to the professors and ask them questions, even me myself as an international student in my first year, I wasn't very outgoing. I was a bit shy.*"

Two issues were mentioned by several students in their answers to different questions: one is the lack of clarity of assignment requirements and assessment criteria, and the other is the lack of training on English academic writing. Interestingly, 15 out of the 25 still preferred writing/reports as the assessment method, in comparison to the alternative audio/video assessment. Some said they were more comfortable with writing, and audio/video assessment might require different skills or more time: "*I don't like writing exams, but I like typing out assignments and submitting it.*"

The degree award gap can vary from subject to subject, which suggests that different solutions (or different combinations of solutions) may be needed for different subjects (Cramer 2021). Cramer (2021) finds the exam is the main contributor to the award gap in the subject of cell biology and points out that comparison of all marks does not reveal any larger award gap in the coursework. In Brunel Design School, the majority of assessment tends to be on coursework, with a few exams. It would be useful to conduct further studies on the gaps between exams and coursework.

The suggestions from the interviews and the student EDI groups were bottom-up, and provide directions for action. However, improving the learning experience of minority ethnic students and addressing the existing award gaps also require top-down strategies. Cramer (2021) proposed five policies and some can be useful for design, e.g.:

1. create time and fund resources to develop, implement and monitor the impact of recommendations;
2. increase scientific evidence on causes of the unexplained award gap and the diversity of approaches;
3. increase direct funding to research teams proposing scientific approaches.

Wong et al. (2021) proposes the following strategies based on his two-year study addressing the degree award gaps at the University of Reading, which is useful for Brunel University London as a reference.

- provide greater economic support for minority ethnic students;
- establish an institutional commitment to challenge all forms of racism;
- increase representation of minority ethnic staff and students in HE;
- diversify the curriculum, incorporating a wider range of values and perspectives;
- ensure structural barriers are reduced, e.g., through alternative provisions.

Focusing on arts and design, Partington (2019) suggests 'a wider, more pluralised, range of ways of learning need to be encouraged and accommodated, to reflect a much more heterogenous mix of students, and the diversity of contemporary culture', this was echoed by our student EDI groups. The concept of diversity can have the effect of sustaining inequalities (Brooks 2020). While the 'heterogenous mix of students' in the UK context often refers to students from different backgrounds (e.g., mature students), it is important to consider minority ethnic students and international students. The English language support could be better designed to address their language and cultural barriers.

5 Actions

A number of actions have been taken at Brunel Design School since March 2022 to improve the minority ethnic students' learning experience:

1. Debriefing: the student participants were invited for a debriefing picnic;
2. Dissemination: the findings were shared at the Brunel Teaching Festival and the Horizon in STEM conference;
3. All assessment briefs were required to be written up and shared with students well in advance;
4. Greater care was taken when organising social events (e.g., inclusive activities, food and drinks);
5. Lecturers were encouraged to incorporate content reflecting different values and perspectives;
6. Inclusivity and care for minority ethnic students were discussed at staff meetings, with actions to follow up;
7. Internship opportunities were created at the School for peer-support;
8. Staff were supported to attend relevant training;
9. Staff-student partnerships were emphasised at Student Experience Committee;
10. Made in Brunel team became more diverse, and they were recommended to talk about their experience at a national HE education conference;
11. A tailored guidance for report writing was provided in Chinese (alternative provision);
12. 3D printing workshops were made available to Post Graduate students;
13. More guest lectures were invited;
14. Enhancing the learning environment with work from a minority ethnic artist.

More actions were planned for the future, for example:

- making coursework markings anonymous (exams are already anonymous);
- providing opportunities for minority ethnic students to have tutors of similar backgrounds;
- providing bridging courses;
- sharing good practice among staff members;
- carefully allocating supervisors for PG students;
- discussing with the Language Centre to improve their support
- comparing the impact of exams on award gaps.

6 Conclusion

To conclude, through interviewing 25 minority ethnic students at Brunel Design School, we have identified a number of areas to improve their learning experience.

It was found that English language skills play an important role in the learning experience and assessment outcomes of minority ethnic students (especially for our Chinese students). Language support can be tailored to their needs. Clarifying requirements and assessment criteria could also help with their academic performance. Despite the lack of training in English academic writing, most of the participants preferred reports rather than video/audio assessment. We will use the students' suggestions and relevant strategies to address award gaps in design. We have taken actions to address some of the issues and planned to do more. These actions have the potential to narrow future award gaps; with the inclusive design principle, they will enhance the learning experience of all our students.

Acknowledgements. The authors thank Aymane Sahli for helping transcribe some of the interviews; Dr Busayawan Lam, Steve Cockett, Steve McGonigal, Alan Mannion, Paul Josse, Rebecca Mellor, Kaelum Van Der Berg, and May Ye for useful discussions and assistance, and all the student volunteers for their participation. The project was funded by Brunel University London.

References

Advance HE: Athena Swan Charter: Encouraging and recognising commitment to advancing gender equality (2020). www.advance-he.ac.uk/equality-charters/athena-swan-charter. Accessed 7 Sept 2022

Advance HE: Advance HE strategy 2021–2024 (2022). www.advance-he.ac.uk/about-us/advance-he-strategy-2021-2024. Accessed 11 Dec 2022

Brooks, R.: Diversity and the European higher education student: policy influencers' narratives of difference. Stud. High. Educ. **45**(7), 1507–1518 (2020). https://doi.org/10.1080/03075079.2018.1564263

BSI: BS 7000-6:2005 Design management systems. Managing inclusive design. Guide. The British Standards Institution (2005)

Loehwing, M., Craig, B.: Diversity to dignity: overcoming polarization in an age of division. J. Scholarsh. Teach. Learn. Indianapolis **21**(4), 11–19 (2021). https://doi.org/10.14434/josotl.v21i4.33216

Mertler, C.A.: Action Research: Improving Schools and Empowering Educators, 5th edn. SAGE Publications Inc., Thousand Oaks (2017)

Minocha, S.: Scholarship of teaching and learning in STEM. Open University free course (2021). www.open.edu/openlearn/science-maths-technology/scholarship-teaching-and-learning-stem/content-section-overview?active-tab=description-tab. Accessed 7 Sept 2022

Partington, A.: Learner or consumer? Market-led higher education, diversity, inclusion, and equality. Res. Educ. Adm. Leadersh. **4**(3), 527–552 (2019)

Scott, C.: Managing and regulating commitments to equality, diversity and inclusion in higher education. Irish Educ. Stud. **39**(2), 175–191 (2020). https://doi.org/10.1080/03323315.2020.1754879

Wolbring, G., Lillywhite, A.: Equity/equality, diversity, and inclusion (EDI) in universities: the case of disabled people. Societies **11**(2), 49 (2021). https://doi.org/10.3390/soc11020049

Wong, B., ElMorally, R., Copsey-Blake, M.: 'Fair and square': what do students think about the ethnicity degree awarding gap? J. Furth. High. Educ. **45**(8), 1147–1161 (2021). https://doi.org/10.1080/0309877X.2021.1932773

Check for updates

An Educational Game for Inclusive Innovation: Teaching How to Design for and with Extreme Users

E. Peyrard(✉)

i3-CRG, UMR 9217 CNRS, École Polytechnique, Palaiseau, France
estelle.peyrard@polytechnique.edu

Abstract. The design of products has an impact on the quality of life of people with disabilities. They can indeed amplify or attenuate the situation of disability. For example, a smartphone can serve as an environmental control or as an alternative communication tool but it can only be used by people with severe motor limitations if it is compatible with assistive technologies. Inclusive innovation is a process that involves excluded groups, such as people with disabilities or elderly, and improves their quality of life and inclusion into society. One of the major challenges of inclusive innovation is to raise the awareness in those who design innovation. Gamification has proved to foster learners' engagement. For this reason we developed an Inclusive Innovation card game aimed at training students and professionals (designers, engineers, entrepreneurs, marketing, living labs, ...) in the Inclusive Innovation process. In this article, we describe the card game and its implementation within student and professional contexts.

Keywords: Inclusive innovation · Education · Serious game · Inclusive design · Cocreation

1 Introduction

It is a well-accepted fact that the environment has an impact on the creation of the situation of disability. Cities, buildings, equipment, objects and consumer goods can increase or attenuate the disability, depending on their accessibility and quality of use. Including people with disabilities or elders in design processes can help to increase accessibility and limit situations of disability.

Inclusive innovation is a process that includes people with disabilities, elders and also persons in precarious situations, to improve their quality of life (George et al. 2012; Schillo and Robinson 2017; Peyrard 2022). Inclusive innovation adopts the objectives of inclusive design, which are to include the whole population through design (Clarkson et al. 2003), and adds the objective of participation of excluded groups in the design process. The participation of excluded groups is required to succeed in inclusive design projects (e.g., Avila et al. 2016; Ramster and Keren 2020). One of the challenges of inclusive innovation is that of convincing those who make the innovation on the question

J. Goodman-Deane et al. (Eds.): CWUAAT 2023, *Design for Sustainable Inclusion*, pp. 210–222, 2023.
https://doi.org/10.1007/978-3-031-28528-8_22

of inclusion. To tackle this challenge, with a research-action methodology, we created a card game aiming at training students, designers and engineers in inclusive innovation.

In this paper, after positioning inclusive innovation and gamification in educational experiences, we present the card game and the way it can be used with students and professionals.

2 Literature Review

2.1 Inclusive Innovation

Inclusive innovation is defined as innovation addressing the needs of disenfranchised people to enhance their quality of life (George et al. 2012). Those include people with disabilities, people in insecure situations, people living in developing countries, women, young people, etc. (Schillo and Robinson 2017). In the domain of disabilities, **inclusive innovation bridges inclusive design and codesign** (Peyrard 2022).

Inclusive design is a design philosophy and set of methods that assume it is possible to design environments and products that suit a wide range of users' needs, including the needs of people with impairments (Story et al. 1998; Clarkson et al. 2003). Universal design principles are far from being widely adopted by companies who still consider it mainly as a cost (Vanderheiden and Jordan 2012).

On the contrary, codesign methodologies are widespread in companies. Derived from user centered design, that put the user's needs at the center of the innovation process (Norman and Draper 1986), codesign requires the involvement of users in the innovation process (Prahalad and Ramaswamy 2004; Sanders and Stappers 2008). It enables companies to better address the needs of their users. Codesign with people with disabilities and the elderly has further specific challenges, among which are:

- Practical challenges, such as enabling users with reduced mobility to participate (Brule et al. 2016),
- Communication challenges when the impairment complicates the expression of requirements (Astell et al. 2009; Magnier et al. 2021; Peyrard and Chamaret 2020; Cappelen and Andersson 2012),
- Methodological challenges related to the capacity of abstraction and problem solving needed for codesign (Hendriks et al. 2015).

2.2 Gamification in Educational Practices

Gamification is the use of gaming techniques in non-game contexts: it is used to foster the engagement of learners and to increase their motivation (da Rocha Seixas et al. 2016). Gamification has been used to raise awareness of various societal topics such as sustainable development (Genevois and Leininger-Frézal 2010) or cyber threats (Cook et al. 2016). Its use various forms include most commonly board or card games and video games.

In the domain of social innovation, various card games have been developed. Friedman and Hendry (2012) have developed *Envisioning cards*, which questions the human values of a design. Viera and colleagues (2020) created an *Inclusive design toolkit* gathering human-centered, gender-centered and inclusive design approaches. Torrens et al. (2020) created a board game for inclusive interior design teaching. Indean, a design subsidiary of Cap Gemini, created *Cards for humanity*[1] that combine personality traits and disabilities to question the answer to various users' needs. The University of Lincoln created *Inclusive signs*[2] (Rossi 2022) that help designers creating meta-design concepts and insights related to inclusion. The University of Cambridge created an *Inclusive design toolkit* (Clarkson et al. 2007), including sets of personas and an exclusion calculator[3]. Yet none of these games addresses the full scope of inclusive innovation, meaning innovation for and with any excluded users. The cocreation part of inclusive innovation, in particular, is not addressed.

3 Methodology

The present research was initiated at the TechLab, the technological department of APF France handicap. APF France handicap is one of the largest French associations for people with disabilities. I was employed by the association create a support service for co-design approaches with people with disabilities, with an action-research setting. At the TechLab, we organise design workshops and product testing with people with disabilities on behalf of startups and large companies who want their products to meet the needs of people with disabilities. 80% of these products are assistive technologies, and 20% are mainstream products.

We led more than 30 cocreation projects between 2018 and 2022. It meant gathering a panel, defining the cocreation methodology, conducting the workshop and analysing the data. Our panels were mainly composed of people with motor disabilities. Some also involved people with cognitive and visual impairments and older people. We also led 5 courses on inclusive innovation for students (engineers and business students) and professionals.

Based on these experiences, on the 4 years of research led at the TechLab and on a literature review, we gathered:

- A set of questions to analyse the exclusion created by a product,
- A set of questions based on universal design principles,
- A set of cocreation methodologies,
- A set of adaptations to include people with any impairment or to address any specific need in panels,
- A set of ethical questions for cocreation contexts.

This led me to create a card game to train students and professionals in inclusive innovation. I tested the game on four groups (Table 1): a multi-competence team dedicated to innovation in the disabilities field (TechLab), a group of social entrepreneurs in a startup acceleration program (21), a group of professionals and students in a meeting dedicated

[1] https://cardsforhumanity.idean.com/.

[2] https://inclusivesigns.lincoln.ac.uk/.

[3] https://www.inclusivedesigntoolkit.com/.

to innovation (Jour i, Angers Technopole) and students in a marketing master's degree (IAE of Nancy). Each test lasted between 2 and 3 h. The tests at Angers Technopole and IAE of Nancy focused on inclusive design (and did not address cocreation). A total of 87 persons involved in testing the game.

I followed an action research methodology, frequently used in social research, involving practitioners in testing the game. Action research combines action and research, arguing that "a social situation can best be understood if a change is introduced into it and its effects are observed" (Lewin 1946; Babüroglu and Ravn 1992). I observed the testers during their group discussions and each session of the game was followed by a discussion with the groups to assess its impact and to identify improvements. Periods of reflection were held between sessions to improve the game and the game play, in order to reinforce the engagement of the players and their understanding of inclusive innovation meanings and practice.

4 The Inclusive Innovation Card Game

Inclusive innovation addresses two different objectives: (1) Innovating for excluded groups, such as people with disabilities, to improve their quality of life and (2) enabling the participation of excluded groups in the innovation process. Six types of cards were created:

- *objects and services* cards,
- *inclusive design cards* that gather
  - *extreme user* cards and
  - *designing for all* cards,
- and *cocreation cards* that gather
  - *cocreation methodology* cards,
  - *involving everyone* cards,
  - *ethics of cocreation* cards.

4.1 Object and Service Cards

The card game can be used either by students or by professionals who are in charge of the design of a product or services (designers, engineers, ...). When tested at 21 and Angers Technopole, the game focused on the products or service of the participants. With the IAE students, "object cards" and "service cards" were randomly chosen for each group to apply the different questions to a particular product or service. Seven object cards were proposed: shoes, coffee maker, salad spinner, sink, smartphone, micro-wave oven, desk and make-up. Four service cards were proposed: a hotel, a bank, a museum and a train station.

Table 1. Testers of the game

Place of the test	Context	No. of participants	Profiles	Date
TechLab, Paris	Workshop to develop the game	9	Designer, innovation project manager, entrepreneurship program leader, engineer, ergotherapist, speech therapist	28/06/22
21 (French Red Cross and Nexem innovation accelerator), Paris area	Workshop of the accelerator program	18	Social entrepreneurs and innovation program leaders	29/09/22
Jour I (Angers Technopole)	Workshop during an event dedicated to innovation, within a technology park	40	People working on R&D positions, entrepreneurs, students, professors	18/10/22
IAE Nancy	Closing of a marketing course	20	Master's students	24/10/22

Some of these examples were chosen because corresponding universal design projects exist such as the Oxo salad spinner and the Tefal Includeo coffee maker that can be presented to the players after the game. Participants are encouraged to take into account the full usage sequence of the product or service, from the information on the product, the purchase, the installation, to the use and maintenance of it (Fig. 1).

Fig. 1. An example of an "Object" card and the Includeo coffee maker by Tefal to serve as an illustration

4.2 Inclusive Design Cards

Inclusive design cards draw attention to two issues: the exclusion created by design and the principles of inclusive design.

The first category of cards, initially named *"all users"* questions the possibility of using a product in some situations and with some impairments. These cards were renamed "*extreme users*" because the term "extreme user" draws attention to the potential of people with disabilities in the design of common goods. The cards present eight domains: vision, hearing, motricity, cognition, communication, morphology (age, size, weight and sex) and precarity (equipment and social condition). After the tests, I followed suggestions from participants to add a personal life category (family configuration, diet, ...) and a chronic disease. For each disability domain, situational disabilities (Story et al. 1998), such as being dazzled by the sun or having one hand occupied, and permanent disabilities are questioned. Some figures enable the player to measure the design exclusion. After the first two tests, I transformed the cards presenting a full category (ex: all disabilities relating to motricity) to smaller cards each presenting a disability (ex: "I have one leg or one arm missing" or "my two hands are busy"). It helped the participants to really question the use of a product or service by each "extreme user". The game now has 70 "extreme users" cards.

The second category of cards, named "*designing for all*", briefly presents the 7 universal design principles, established by the Center for Universal Design in the 1990s: Equitable Use, Flexibility in Use, Simple and Intuitive Use, Perceptible Information, Tolerance for Error, Low Physical Effort and Size and Space for Approach and Use (Story et al. 1998). After the first tests, I also transformed the large cards presenting the full principles (ex: flexibility in use) into small cards providing one piece of advice related to one principle (ex: are there several command modes for your product?). Focusing on one question at a time helped the player. The game now has 48 "designing for all" cards.

4.3 Cocreation Cards

Cocreation cards aim at helping students and practitioners to build their cocreation sessions. Three types of cards are proposed: cocreation method, involving everyone and ethics of cocreation.

The *"cocreation method"* cards describe six standard methods for cocreation: user interviews, user observation, auto-observation, quantitative study, focus groups and user testing. We insist on the preparation phase of the cocreation (definition of the objectives, test of the methodology) and on the analysis phase. These methods are classical methods used by designers. Yet the test of the game with social entrepreneurs showed that these methods are not widespread within startups ecosystems. It also underlined the need to insist on the posture of the entrepreneurs toward participants in cocreation and on the possible bias associated to this posture.

The *"involving everyone"* cards provide guidelines for involving persons with all types of impairments and situations. They first describe the impact that an impairment can have during a cocreation session. They then make recommendations on adaptations needed for the cocreation session to be inclusive. They address the following categories of impairments and situations: motor impairment, visual impairment, hearing impairment, mental and cognitive impairment, communication and behavioral impairment, aged and young participants, participants with various social backgrounds. The adaptations derive from the methodology built at the TechLab (Peyrard 2022) and other projects mentioned in the literature (Orpwood et al. 2005; Plos 2011; Magnier et al. 2012; Guffroy 2017) (Figs. 2 and 3).

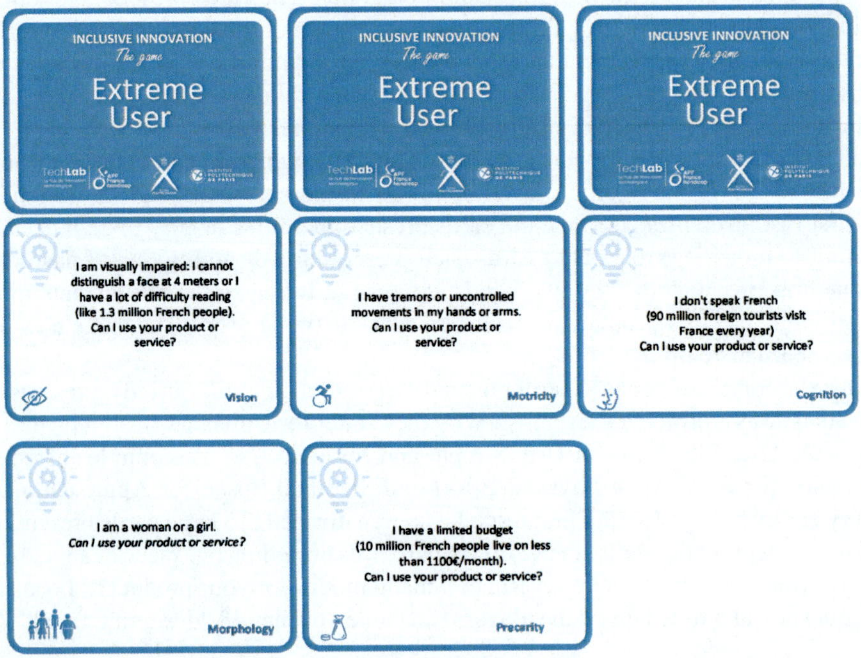

Fig. 2. Five examples of the "Extreme users" cards that question the exclusion created by design

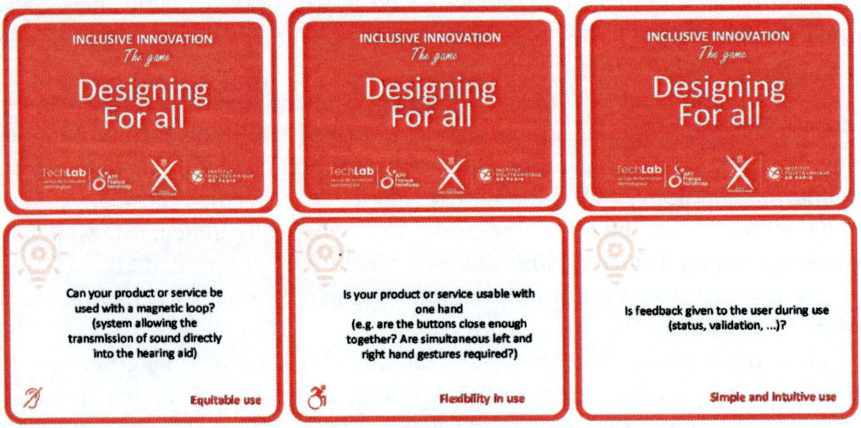

Fig. 3. Three examples of the "Designing for all" card that question the application of universal design principles

The "*ethics of cocreation*" cards provide questions on 8 ethical dimensions of cocreation: consent, openness, personal data, security, expectations sharing, value sharing, confidentiality and influence. These are directly derived from the research led at the TechLab and the analysis of the projects led (Peyrard 2022).

Our tests showed that the cocreation cards do not address exactly the same public as inclusive design cards. Inclusive design cards can indeed be played with novices, with any person interested in innovation. They help in raising awareness on the exclusion created *by design* and give paths to apply inclusive design. Cocreation cards address people that are already in an inclusive mindset, ready to involve people with disabilities in their cocreation process. They help in creating a cocreation methodology. They raise awareness of the methodological and ethical stakes. With students, the two card games could be used to teach the two parts of inclusive innovation: inclusive design and cocreation (Figs. 4, 5 and 6).

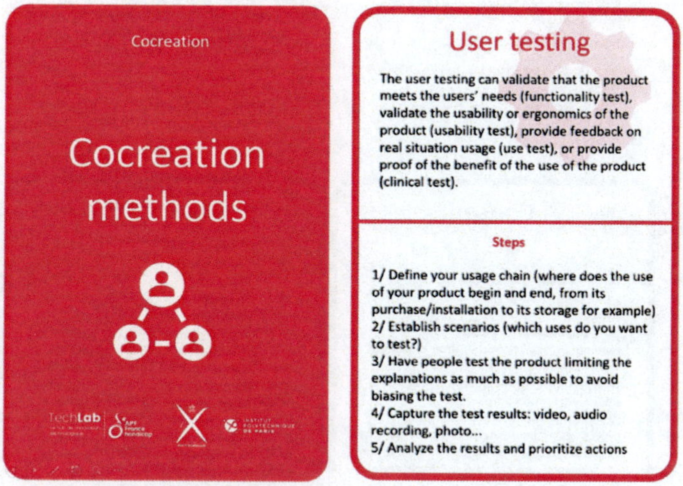

Fig. 4. One example of "cocreation methods" card

5 A Game Design Embracing the Inclusive Innovation Process

The gameplays evolved after each of the four tests. Three gameplays emerged, one for players developing a product or service, one for players not developing a product or service and one for players already engaged in a social and/or inclusive innovation project.

5.1 Training Players Not Developing a Product or Service

In this gameplay, the players do not focus on their own product or service but focus on products or services that we suggest. They are asked to form groups of 3 or 4 people. This gameplay focus on the "inclusive design" part of the game. The main goal is thus to teach the participants how to design a better, more inclusive, product or service. The game consists of several parts:

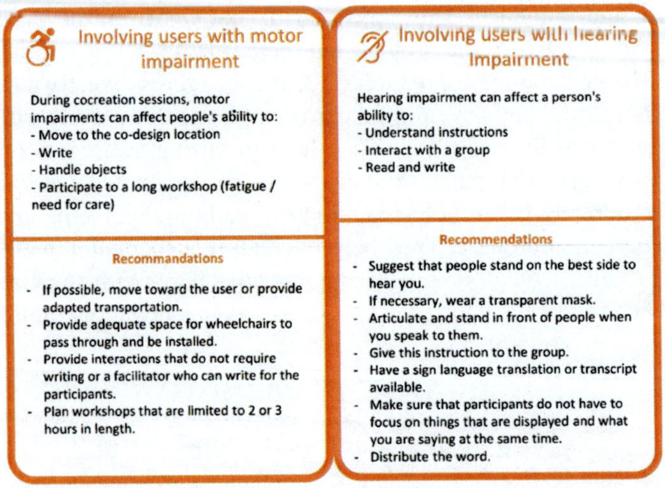

Fig. 5. Two examples of "*Involving everyone*" cards

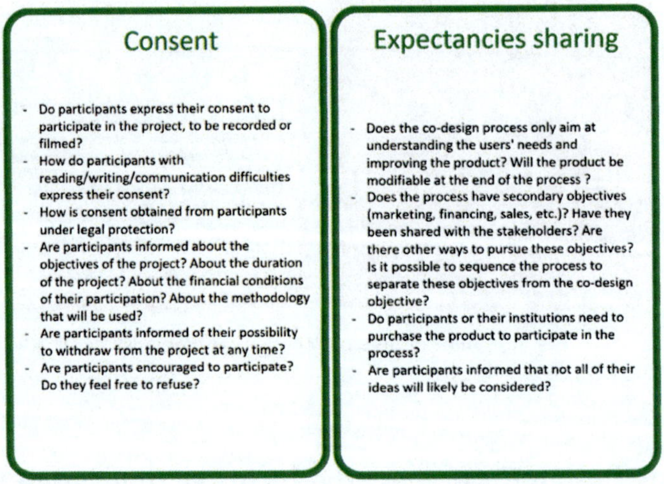

Fig. 6. Two examples of "*Ethics of cocreation*" cards

1. The game starts with an introduction of design exclusion and extreme users. It shows several examples of products invented or improved for the needs of extreme users (ex: the remote control).
2. An "object" or "service" card is given to each group.
3. "*Extreme users*" full set of cards are given to each group. Within a defined time (30 min in our tests), they are asked to identify exclusions created by the design of their object or service and resolution ideas (either improvement of the object or a totally new object). They are invited to consider the whole usage sequence of the product or service (information, purchase, installation, use, …) and to reflect on how the improvement for extreme users will benefits all users. To foster participants'

engagement in the game, they could be asked to find as many exclusions as possible (1 point) and as many resolutions as possible (2 points).
4. Each group shares the exclusions and resolutions identified with the class.
5. A presentation of the 7 principles of inclusive design is made to the class.
6. *"design for all"* full sets of cards are given to each group. Within a defined time (20 min in our tests), they are asked to imagine new improvements to their object or service. To foster learning and avoid repetitions, a new "object" or "service" card could be given to each group. Engagement could also be fostered at this stage with a points system (2 points per new improvement).
7. Again, each group shares the improvements they thought of.
8. The game can stop here with presentations of products adapted by famous brands (ex: Nike Flyease or Tefal Includeo). In our tests, these 7 steps took 2 to 3 h with 20 to 50 participants.
9. If the course is longer, the participants can then be asked to go deeper in the product or service definition: product visual, product characteristics, prototype, business plan, marketing plan…

An option would be to ask some groups to work on the worst product in terms of inclusiveness.

5.2 Training Players Developing a Product or Service

The card game can also be used with professionals such as entrepreneurs, designers, engineers or members of the marketing team actually developing a product or service. The gameplay is overall the same but the focus can be a little bit different.

In the "extreme user" phase, participants can be asked to split the *extreme users* into 4 categories: non-extreme user (user who can easily use the product or service), extreme user (user who can use the product or service with difficulties), non-user (user who is excluded from the use of the product or service), out of scope user (user who can logically not use the product or service – ex: a man cannot use bras). The teacher can review and challenge the "out of scope users".

In the "design for all" phase, participants could select a few extreme users and non-users to work on. The icons on the "design for all" cards will help them identify the best practices to addressing different types of extreme users.

In the conclusion phase, participants can be asked to place each improvement they identified on a matrix "easy vs difficult to realise/small vs big impact". An introduction to cocreation with extreme users can also be done.

5.3 Training Players Engaged in a Social and/or Inclusive Innovation Project

With players already engaged in a social and/or inclusive innovation project, the first phase can be played as described in the previous paragraph. A second phase of the teaching is aimed at organising a cocreation session with excluded users. Within a 2 h workshop, if each participant is engaged in a different project, the different steps can be:

1. Each participant goes through the set of *«cocreation methods»* cards to select one or two methods. Most entrepreneurs had already interacted with users but they did not have knowledge of the different possible types of interactions.
2. Each participant shares its choices with the group.
3. Each participant defines its ideal panel and its cocreation methodology and adaptations with the set of "*involving everyone*" cards. The attention of participants had to be focused on the methodology of cocreation because the participants tended to use disabilities descriptions to think again to product improvements. It suggests that for already engaged entrepreneurs or designers, a presentation of the different types of disabilities could complete the first part of the game.
4. Each participant shares its methodology with the group.
5. Each participant chooses two "*ethical questions*" cards that they have not already taken into account.
6. Each participant shares its thoughts with the group.

According to the time available for the teaching, the participants can lead their cocreation sessions, analyse the results, and adapt their product definition.

6 Conclusion

We developed a card game as a tool to build engagement in inclusive innovation teaching. The originality of the game is to embrace a full range of *extreme users*, to make participants work on the innovation potential of those users and to address cocreation methodologies with *extreme users*. It can be used with students and professionals and brings them to consider the universality of a design and methods to involve people with disabilities in their innovation process. Our first tests and participants feedbacks showed that within a 2- or 3-hour workshop, reflecting on extreme users and universal design approaches enables participants not only to identify improvements on any product or service but also to acquire a culture of inclusive innovation. Further research will be needed to measure the exact impact of this game, both on the creativity of companies and on the culture change toward more inclusive ways to design products and services. The game still has to be tested as a support for longer teaching.

Acknowledgement. The research led at the TechLab of APF France handicap received support from Mutuelle Intégrance, Harmony Mutuelle ESS and Société Générale.

References

Astell, A., Alm, N., Gowans, G., Ellis, M., Dye, R., et al.: Involving older people with dementia and their carers in designing computer based support systems: some methodological considerations. Univ. Access Inf. Soc. **8**, 49–58 (2009). https://doi.org/10.1007/s10209-008-0129-9

Avila, C., Baldiris, S., Fabregat, R., Graf, S.: Cocreation and evaluation of inclusive and accessible open educational resources: a mapping toward the IMS caliper. IEEE Revista Iberoamericana de Tecnologias del Aprendizaje **11**(3), 167–176 (2016). https://doi.org/10.1109/RITA.2016.2589578

Babüroglu, O.N., Ravn, I.: Normative action research. Organ. Stud. **13**(1), 019–034 (1992). https://doi.org/10.1177/017084069201300104

Brule, E., Bailly, G., Brock, A., Valentin, F., Denis, G., et al.: MapSense: multi-sensory interactive maps for children living with visual impairments. In: Proceedings of the 2016 CHI Conference on Human Factors in Computing Systems (2016). https://doi.org/10.1145/2858036.2858375

Cappelen, B., Andersson, A.-P.: Trans-create–co-design with persons with severe disabilities. In: Universal Design 202: From Special to Mainstream Solutions, vol. 282, pp. 87–101. IOS Press (2021). https://doi.org/10.3233/SHTI210387

Clarkson, J., Coleman, R., Hosking, I., Waller, S.: Inclusive Design Toolkit. Cambridge Engineering Design Centre, UK (2007)

Clarkson, J., Keates, S., Coleman, R., Lebbon, C.: Inclusive Design: Design for the Whole Population. Springer, Cham (2003). https://doi.org/10.1007/978-1-4471-0001-0

Cook, A., Smith, R., Maglaras, L., Janicke, H.: Using gamification to raise awareness of cyber threats to critical national infrastructure. In: Proceedings of the 4th International Symposium for ICS & SCADA Cyber Security Research, Belfast, pp. 84–94, August 2016

da Rocha, S.L., Gomes, A.S., de Melo Filho, I.J.: Effectiveness of gamification in the engagement of students. Comput. Hum. Behav. **58**, 48–63 (2016). https://doi.org/10.1016/j.chb.2015.11.021

Friedman, B., Hendry, D.: The envisioning cards: a toolkit for catalyzing humanistic and technical imaginations. In: CHI 2012: Proceedings of the SIGCHI Conference on Human Factors in Computing Systems, May 2012, pp. 1145–1148 (2012). https://doi.org/10.1145/2207676.2208562

Genevois, S., Leininger-Frézal, C.: Les “serious games”: Un outil d’éducation au développement durable? Colloque International “Education au développement durable et à la biodiversité: Concepts, questions vives, outils et pratiques”, Digne les Bains, France (2010)

George, G., McGahan, A.M., Prabhu, J.: Innovation for inclusive growth: towards a theoretical framework and a research agenda. J. Manage. Stud. **49**(4), 661–683 (2012). https://doi.org/10.1111/j.1467-6486.2012.01048.x

Guffroy, M.: Adaptation de méthodes d’évaluation dans le cadre de la conception d’une application numérique pour un jeune public avec troubles du spectre autistique : Étude au cours de la conception et de l’évaluation de l’application çATED au sein d’une ULIS TED. PhD Theses, Le Mans (2017). www.theses.fr/2017LEMA3001. Accessed 25 Jan 2023

Hendriks, N., Slegers, K., Duysburgh, P.: Codesign with people living with cognitive or sensory impairments: a case for method stories and uniqueness. CoDesign **11**(1), 70–82 (2015). https://doi.org/10.1080/15710882.2015.1020316

Lewin, K.: Action research and minority problems. J. Soc. Issues **2**(4), 34–46 (1946). https://doi.org/10.1111/j.1540-4560.1946.tb02295.x

Magnier, C., Thomann, G., Villeneuve, F., Zwolinski, P.: Methods for designing assistive devices extracted from 16 case studies in the literature. Int. J. Interact. Des. Manuf. **6**(2), 93–100 (2012). https://doi.org/10.1007/s12008-012-0143-2

Norman, D.A., Draper, S.W.: User Centered System Design. New Perspectives on Human-Computer Interaction, 1st edn. CRC Press, Boca Raton (1986)

Orpwood, R.D., Gibbs, C., Adlam, T.D., Faulkner, R., Meegahawatte, D.: The design of smart homes for people with dementia—user-interface aspects. Univ. Access Inf. Soc. **4**(2), 156–164 (2005). https://doi.org/10.1007/s10209-005-0120-7

Peyrard, E.: L’innovation inclusive en pratique: Fondations, apports et défis dans le domaine du handicap. PhD theses, Institut polytechnique de Paris, France (2022). www.theses.fr/2022IPPAX015. Accessed 16 Jan 2023

Peyrard, E., Chamaret, C.: Concevoir pour tous, mais avec qui? Trois cas de co-conception avec des personnes en situation de handicap. Gérer et Comprendre. Annales des Mines **141**(3), 57–70 (2020)

Plos, O.: Innover pour et par le handicap méthodologie de conception de produits adaptée aux marchés de niche: Application au marché du handicap moteur. PhD theses, Arts et Métiers ParisTech, France (2011). hal.archives-ouvertes.fr/pastel-00578706/document. Accessed 16 Jan 2023

Prahalad, C.K., Ramaswamy, V.: Co-creating unique value with customers. Strategy Leadership **32**(3), 4–9 (2004). https://doi.org/10.1108/10878570410699249

Ramster, G., Keren, C.: Designing with communities: lessons on social design from the Helen Hamlyn Centre for Design. Royal College of Art, Postgraduate Art & Design, December 2020, London, UK (2020)

Rossi, E.: Inclusive signs: a card-based toolkit to generate creative inclusive design concepts and research strategies. University of Lincoln, UK (2022)

Sanders, E.B.-N., Stappers, P.J.: Co-creation and the new landscapes of design. CoDesign **4**(1), 5–18 (2008). https://doi.org/10.1080/15710880701875068

Schillo, R.S., Robinson, R.: Inclusive innovation in developed countries: the who, what, why, and how. Technol. Innov. Manag. Rev. **7**(7), 34–46 (2017). https://doi.org/10.22215/timreview/1089

Story, M.F., Mueller, J.L., Mace, R.L.: The universal design file: Designing for people of all ages and abilities. Revised Edition. Center for Universal Design, NC State University, USA (1998). eric.ed.gov/?id=ED460554. Accessed 16 Jan 2023

Torrens, G.E., Jiang, Y., Dong, H.: Incorporating "care" into design education through games. In: Tosi, F., Lacono, E., Brischetto, A., Serra, A. (eds.) Design for Inclusion, Gamification and Learning Experience, pp. 54–68. Francoangeli (2020). www.torrossa.com/en/catalog/preview/4925432. Accessed 16 Jan 2023

Vanderheiden, G.C., Jordan, J.B.: Design for people with functional limitations. In: Salvendy, G. (ed.) Handbook of Human Factors and Ergonomics, pp. 1407–1441. Wiley (2012). https://doi.org/10.1002/9781118131350.ch51

Viera, D.F., González-González, C., Iranzo, R.G., Paderewski, P., Gutiérrez Vela, L.G.: Inclusive design toolkit for the creation of intergenerational gamified experiences. In: ACM International Conference Proceeding Series, pp. 707–712 (2020). https://doi.org/10.1145/3434780.3436697

Author Index

J. Goodman-Deane et al. (Eds.): CWUAAT 2023, *Design for Sustainable Inclusion*, pp. 223–224, 2023.
https://doi.org/10.1007/978-3-031-28528-8